James Monteith

Comprehensive Geography

The Independent Course

James Monteith

Comprehensive Geography
The Independent Course

ISBN/EAN: 9783742807380

Manufactured in Europe, USA, Canada, Australia, Japa

Cover: Foto ©berggeist007 / pixelio.de

Manufactured and distributed by brebook publishing software
(www.brebook.com)

James Monteith

Comprehensive Geography

THE INDEPENDENT COURSE.

COMPREHENSIVE

GEOGRAPHY;

LOCAL,
PHYSICAL,
DESCRIPTIVE,
HISTORICAL,
MATHEMATICAL,
COMPARATIVE,
TOPICAL,
and ANCIENT;

WITH

MAP-DRAWING AND RELIEF MAPS.

By JAMES MONTEITH,

AUTHOR OF A SERIES OF GEOGRAPHIES, MAPS AND GLOBES.

A. S. BARNES & COMPANY, PUBLISHERS,

NEW YORK, CHICAGO, AND NEW ORLEANS.

1876.

INDEX TO CONTENTS.

PLAN OF THE COMPREHENSIVE GEOGRAPHY.

THIS work is adapted to every grade, and is prepared with especial reference to conciseness, simplicity, and that diversity of presentation which is so essential in a school-room.

Owing to the limited time which many scholars can give to the study of Geography, the author has endeavored to secure satisfactory results at the smallest possible expense in time and labor, both to the teacher and pupil.

One leading feature is that the student learns all about *one country or state at a time*—its Civil, Physical, Descriptive, Comparative, and Historical Geography; thus enabling him to obtain an uninterrupted view of its geography in its several branches, and to observe the effects of climate and productions upon the conditions and pursuits of the inhabitants.

Each division of the earth's surface is presented, first, in regard to its size, shape, and position on the globe; then general questions on its outline, mountains, rivers, cities, etc., followed by map-drawing (according to a uniform scale) which will impress firmly upon the mind all that has been mentioned in the recitation.

THE RELIEF MAPS, or bird's-eye views of the earth's surface, with exercises, lead the student to *learn from observation* as well as study.

THE HISTORICAL GEOGRAPHY gives a brief outline of each country or state—its settlement, progress, and most celebrated rulers—sufficient to impress upon the mind of the pupil the connection of event with the place of occurrence, and to induce him to pursue the subject by reading works on history exclusively.

COMPARATIVE LATITUDES are shown on the margins of the maps.

COMPARATIVE AREAS are taught by representing various states and countries drawn upon a common standard. That standard is the State of *Kansas*, whose shape and dimensions—an oblong, 200 by 400 miles—are easily remembered.

REFERENCE MAPS of the principal countries in Europe are included, but the exercises refer only to the prominent places; the names in fine lettering being designed only for reference.

To interest scholars in the use of globes, every book is furnished with a full set of map segments, handsomely engraved and colored, from which any scholar may easily construct a globe by following the directions given.

Accompanying the segments are some useful exercises on latitudes, longitudes, positions, etc., which can be understood only by reference to a globe.

ANCIENT GEOGRAPHY is given in a brief sketch, with maps of the Roman Empire, Ancient Italy, and Ancient Greece.

THE TOPICAL GEOGRAPHY on pages 53, 57, 81, and 88, compares the leading characteristics of one country with those of another, and serves as a Review of the previous lessons, and for the examination of classes.

POPULATION OF MORE THAN 400 CITIES, TOWNS, AND VILLAGES IN THE UNITED STATES.

From the Census of 1870. * The star indicates the Population in 1875.

ALABAMA.
Mobile ... 32,034
Montgomery ... 10,588
Selma ... 6,484

ARIZONA.
Tucson ... 3,224

ARKANSAS.
Little Rock ... 12,380
Pine Bluff ... 2,081
Hot Springs ... 1,276
Van Buren ... 985

CALIFORNIA.
S. Francisco ... 149,473
Sacramento ... 16,283
Oakland ... 10,500
Stockton ... 10,066
San Jose ... 9,089
Grass Valley ... 7,063
Vallejo ... 6,391
Los Angeles ... 5,728
Marysville ... 4,738
Petaluma ... 4,588
Nevada ... 3,986

CONNECTICUT.
New Haven ... 50,840
Hartford ... 37,180
Bridgeport ... 18,969
Norwich ... 16,653
Norwalk ... 12,119
Waterbury ... 10,826
Stamford ... 9,714
New Britain ... 9,480

COLORADO.
Denver ... 4,759
Central City ... 2,360

DELAWARE.
Wilmington ... 30,841
Smyrna ... 2,119
New Castle ... 1,916
Dover ... 1,906

DIST. COLUMBIA.
Washington ... 109,199
Georgetown ... 11,384

FLORIDA.
Jacksonville ... *11,000
Pensacola ... *3,347
Tallahassee ... *2,534

GEORGIA.
Savannah ... *30,000
Atlanta ... *30,000
Augusta ... *30,000
Macon ... *13,000
Columbus ... *9,000
Athens ... *6,000
Milledgeville ... *3,500
Rome ... *3,500
Americus ... *5,000

ILLINOIS.
Chicago ... *395,408
Quincy ... *35,670
Peoria ... *26,247
Springfield ... *25,116
Bloomington ... 14,590
Aurora ... 11,162
Rockford ... 11,049
Galesburg ... 10,158
Jacksonville ... 9,203
Alton ... 8,665
Belleville ... 8,146
Rock Island ... *10,814
Freeport ... 7,889
Ottawa ... 7,736
Joliet ... 7,263
Decatur ... 7,161
Galena ... 7,019
Pekin ... 5,696
E. St. Louis ... 5,644
La Salle ... 5,200
Kankakee ... 5,189
Danville ... 4,751
Monmouth ... 4,662
Waukegan ... 4,507
Moline ... 4,166
Dixon ... 4,055
Sterling ... 3,998
Litchfield ... 3,852
Warsaw ... 3,580
Mendota ... 3,546
Canton ... 3,308
Cairo ... *10,974
Elgin ... 5,441

INDIANA.
Indianapolis ... *62,162
Evansville ... 21,830
Fort Wayne ... *25,800
Terre Haute ... *19,724
New Albany ... 15,396
Lafayette ... 13,506
Madison ... *11,778
Richmond ... 9,445
Logansport ... 8,950
Jeffersonville ... 7,254
South Bend ... 7,206
La Porte ... 6,581
Michigan City ... 3,985
Peru ... 3,617
Crawfordsville ... 3,701

IOWA.
Davenport ... *24,840
Dubuque ... *25,127
Burlington ... 14,930
Keokuk ... *14,500
Des Moines ... *16,250
Council Bluffs ... 10,020
Muscatine ... 6,718
Clinton ... 6,129
Cedar Rapids ... 5,940
Iowa City ... 5,914
Ottumwa ... 5,214
Waterloo ... 4,337
Mt. Pleasant ... 4,245
Lyons ... 4,088
Fort Madison ... 4,011
Sioux City ... 3,401
Marshalltown ... 3,218
Oskaloosa ... 3,204
Fort Dodge ... 3,095
Cedar Falls ... 3,070
Independence ... 3,045
Washington ... 2,575
Waverly ... 2,291
Decorah ... 2,110
Anamosa ... 2,083
McGregor ... 2,074
Chariton ... 1,728

KANSAS.
Leavenworth ... 17,873
Lawrence ... 8,320
Atchison ... 7,054
Topeka ... 5,790
Fort Scott ... 4,174
Ottawa ... 2,941
Wyandotte ... 2,940
Emporia ... 2,168

KENTUCKY.
Louisville ... *120,000
Covington ... *24,505
Newport ... 15,087
Lexington ... 14,801
Paducah ... 6,866
Frankfort ... 5,396
Maysville ... 4,705
Bowling Green ... 4,574
Henderson ... 4,171
Owensboro ... 3,437
Versailles ... 3,268
Hopkinsville ... 3,136
Danville ... 2,542
Shelbyville ... 3,150

LOUISIANA.
New Orleans ... *203,439
Baton Rouge ... 6,948
Shreveport ... 4,607
Monroe ... 1,949
Thibodeaux ... 1,922
Donaldsonville ... 1,573
Opelousas ... 1,546
New Iberia ... 1,472
Plaquemines ... 1,460
Natchitoches ... 1,401
Franklin ... 1,365

MAINE.
Portland ... 34,420
Bangor ... 18,289
Lewiston ... 13,600
Biddeford ... 10,282
Augusta ... 7,808
Bath ... 7,371
Rockland ... 7,074
Auburn ... 6,169
Calais ... 5,944
Belfast ... 5,278
Hallowell ... 3,007

MARYLAND.
Baltimore ... 267,354
Frederick ... 8,526
Cumberland ... 8,056
Hagerstown ... 5,779
Annapolis ... 5,744

MASSACHUSETTS.
Boston ... *341,919
Lowell ... *49,657
Worcester ... *49,265
Cambridge ... *47,832
Fall River ... *45,340
Lawrence ... *34,901
Lynn ... *32,000
Springfield ... *31,026
Salem ... *26,053
New Bedford ... *25,876
Somerville ... *21,868
Chelsea ... *20,695
Taunton ... *20,429
Gloucester ... *16,754
Holyoke ... *16,260
Newton ... *16,105
Haverhill ... *14,628
Newburyport ... *13,323
Fitchburg ... *12,289
Pittsfield ... *12,267

MICHIGAN.
Detroit ... *103,770
Grand Rapids ... 16,507
Jackson ... 11,447
E. Saginaw ... 11,350
Adrian ... 8,438
Kalamazoo ... 9,181
Saginaw ... 7,460
Ann Arbor ... 7,363
Bay City ... 7,064
Muskegon ... 6,002
Port Huron ... 5,973
Battle Creek ... 5,838
Ypsilanti ... 5,471
Flint ... 5,386
Lansing ... *7,632
Monroe ... 5,086
Marshall ... 4,925
Pontiac ... 4,867
Niles ... 4,630
Cold Water ... 4,381
Hillsdale ... 3,518
Manistee ... 3,343
Wyandotte ... 2,731
Owasso ... 2,065
St. Clair ... 1,790
La Peer ... 1,772
Big Rapids ... 1,237

MINNESOTA.
St. Paul ... *45,147
Minneapolis ... 13,066
Winona ... *10,743
St. Anthony ... 5,013
Red Wing ... 4,260
Stillwater ... 4,124
Rochester ... 3,953
Mankato ... 3,482
Hastings ... 3,458
Duluth ... 3,131
Faribault ... 3,045
St. Cloud ... 2,161
St. Peter ... 2,124
Owatonna ... 2,070
Austin ... 2,039
Shakopee City ... 1,347

MISSISSIPPI.
Vicksburg ... 12,443
Natchez ... 9,057
Columbus ... 4,812
Jackson ... 4,234
Holly Springs ... 2,406
Aberdeen ... 2,029
Grenada ... 1,887

MISSOURI.
St. Louis ... *414,398
Kansas City ... *34,482
St. Joseph ... *20,577
Hannibal ... 10,125
St. Charles ... 5,570
Springfield ... 5,555
Sedalia ... 4,560
Jefferson City ... 4,420
Lexington ... 4,373
Chillicothe ... 3,978
Macon ... 3,678
Louisiana ... 3,639
Cape Girardeau ... 3,585
Boonville ... 3,506
Independence ... 3,184
Warrensburg ... 2,945
Palmyra ... 2,615
Mexico ... 2,602
Pleasant Hill ... 2,354

NEBRASKA.
Omaha ... 16,083
Nebraska ... 6,050

NEVADA.
Virginia ... 7,048
Gold Hill ... 4,311
Hamilton ... 3,913
Carson City ... 3,042

NEW HAMPSHIRE.
Manchester ... 23,536
Concord ... 12,241
Nashua ... 10,543
Dover ... 9,294
Portsmouth ... 9,211

NEW JERSEY.
Newark ... 105,059
Jersey City ... *110,000
Paterson ... 33,579
Trenton ... 22,874
Elizabeth ... 20,832
Hoboken ... 20,297
Camden ... 20,045
N. Brunswick ... 15,058
Orange ... 9,348
Bridgeton ... 6,830
Rahway ... 6,258
Millville ... 6,101
Burlington ... 5,817
Plainfield ... 5,095
Union ... 4,640
Harrison ... 4,129

NEW YORK.
New York ... *1,249,868
Brooklyn ... *483,252
Buffalo ... *134,238
Albany ... *84,153
Rochester ... *82,000
Syracuse ... *49,808
Troy ... *48,706
Utica ... *32,089
Cohoes ... *23,000
Oswego ... *22,280
Poughkeep. ... *20,087
Elmira ... *20,093
Auburn ... *19,616
Yonkers ... *17,312
Newburg ... 17,041
Binghamton ... *15,486
Schenectady ... *13,171
Lockport ... 12,426
Rome ... 11,000
West Troy ... 10,693
Rondout ... 10,114
Ogdensburg ... 10,076
Watertown ... 9,336
Hudson ... *8,794
Ithaca ... 8,402
Saratoga Spr. ... 7,516
New Brighton ... 7,495
Kingston ... 6,315
Flushing ... 6,223
Long Is. City ... 3,867

NORTH CAROLINA.
Wilmington ... 13,446
Raleigh ... 7,790
Newberne ... 5,849
Fayetteville ... 4,660
Charlotte ... 4,473

OHIO.
Cincinnati ... *266,568
Cleveland ... *145,693
Toledo ... *54,706
Columbus ... *42,707
Dayton ... 30,473
Sandusky ... 13,000
Springfield ... 12,652
Hamilton ... 11,081
Portsmouth ... 10,592
Zanesville ... 10,011
Akron ... 10,006
Chillicothe ... 8,920
Canton ... 8,660
Steubenville ... 8,107
Youngstown ... 8,075
Mansfield ... 8,029
Newark ... 6,698
Xenia ... 6,377
Piqua ... 5,967
Pomeroy ... 5,824
Ironton ... 5,686
Delaware ... 5,641
Fremont ... 5,455
Wooster ... 5,419
Circleville ... 5,407
Marietta ... 5,218

OREGON.
Portland ... 8,293
Oregon City ... 1,382
Dalles ... 942
Forest Grove ... 922
Eugene City ... 861

PENNSYLVANIA.
Philadelphia ... *806,000
Pittsburg ... 86,076
Allegheny ... 53,180
Scranton ... 35,092
Reading ... 33,930
Harrisburg ... 23,104
Lancaster ... 20,233
Erie ... 19,646
Williamsport ... 16,030
Allentown ... 13,884
Pottsville ... 12,384
York ... 11,003
Easton ... 10,987
Norristown ... 10,753
Altoona ... 10,610
Wilkesbarre ... 10,174
Titusville ... 8,639
Danville ... 8,436
Meadville ... 7,103
Lockhaven ... 6,986
Pittston ... 6,760
Lebanon ... 6,727
Carlisle ... 6,650
Columbia ... 6,461
Carbondale ... 6,393
Chambersburg ... 6,308
New Castle ... 6,164
Johnstown ... 6,028
Tamaqua ... 5,960
St. Clair ... 5,726
Ashland ... 5,714
Westchester ... 5,630
Mahanoy ... 5,533

RHODE ISLAND.
Providence ... *100,675
Newport ... 12,521

SOUTH CAROLINA.
Charleston ... 48,956
Columbia ... 9,298
Greenville ... 2,757
Georgetown ... 2,080

TENNESSEE.
Memphis ... 40,226
Nashville ... *37,242
Knoxville ... 8,682
Chattanooga ... 6,093
Jackson ... 4,119
Murfreesboro ... 3,502
Clarksville ... 3,200
Columbia ... 2,550
Brownsville ... 2,457

TEXAS.
Galveston ... 13,818
San Antonio ... 12,256
Houston ... 9,382
Brownsville ... 4,905
Jefferson ... 4,190
Victoria ... 2,534
Brenham ... 2,221
Laredo ... 2,046
Austin ... 4,428

UTAH.
Salt Lake City ... 12,854
Ogden ... 3,127
Provo ... 2,384
Logan ... 1,757
Mt. Pleasant ... 1,346
Brigham City ... 1,315

VERMONT.
Burlington ... 14,387
Rutland ... 9,834
Bennington ... 2,501
Vergennes ... 1,570
Brattleboro ... 4,933

VIRGINIA.
Richmond ... 51,038
Norfolk ... 19,229
Petersburg ... 18,950
Alexandria ... 13,570
Portsmouth ... 10,492
Lynchburg ... 6,825
Winchester ... 4,477
Fredericksburg ... 4,046
Danville ... 3,463
Lexington ... 2,873
Charlottesville ... 2,838
Manchester ... 2,599
Hampton ... 2,300
Harrisonburg ... 2,036

WASHINGTON TER.
Walla Walla ... 1,394
Olympia ... 1,203
Seattle ... 1,107

WEST VIRGINIA.
Wheeling ... 19,280
Parkersburg ... 5,546
Martinsburg ... 4,863
Charleston ... 3,162
Grafton ... 1,967
Moundsville ... 1,509

WISCONSIN.
Milwaukee ... *100,775
Fond du Lac ... 12,764
Oshkosh ... 12,663
Racine ... 9,880
Madison ... 9,176
Janesville ... 8,789
La Crosse ... *10,596
Watertown ... 7,550
Sheboygan ... 5,310
Manitowoc ... 5,168
Green Bay ... 4,666
Appleton ... 4,518
Beloit ... 4,396
Kenosha ... 4,309
Portage ... 3,945
Monroe ... 3,408
Mineral Point ... 3,275
Beaver Dam ... 3,265
Ripon ... 2,976
Berlin ... 2,777
Prairie du Chien ... 2,700
Neenah ... 2,655
Oconto ... 2,655
Waukesha ... 2,633

AGRICULTURAL PRODUCTS OF THE UNITED STATES, 1870.

STATES & TERRITORIES.	Area in sq. miles.	Population 1870.	Wheat.	Rye.	Corn.	Oats.	Potatoes, white.	Potatoes, sweet.	Tobacco.	Cotton.	Wool.	Butter.	Hay.
			Bushels.	*Bushels.*	*Bushels.*	*Bushels.*	*Bushels.*	*Bushels.*	*Pounds.*	*Bales.*	*Pounds.*	*Pounds.*	*Tons.*
Total United States	3,603,884	38,925,598	287,067,856	16,918,795	760,944,549	282,107,157	143,337,473	21,709,824	262,735,341	3,011,996	100,102,387	514,092,683	27,316,048
Alabama	50,722	996,992	1,055,068	18,977	16,977,948	770,866	162,512	1,871,360	152,742	429,482	381,253	3,213,753	10,613
Arkansas	52,198	484,471	741,736	27,645	13,382,145	328,777	422,196	896,531	594,886	247,968	214,784	2,753,931	6,830
California	188,981	560,247	16,676,702	26,275	1,221,222	1,757,507	2,049,227	263,655	63,809	34	11,391,743	7,969,744	551,773
Connecticut	4,750	537,454	38,144	289,031	1,570,364	1,114,595	2,789,894	867	8,328,798	....	254,129	6,716,007	563,328
Delaware	2,120	125,015	895,477	10,222	3,010,390	554,688	362,724	85,900	250	....	58,216	1,171,963	41,890
Florida	59,268	187,748	....	345	2,225,056	114,204	10,218	789,456	157,405	39,789	37,562	100,989	17
Georgia	58,000	1,184,109	2,127,017	82,549	17,646,459	1,904,601	137,101	2,621,562	288,596	473,934	846,947	4,499,579	10,518
Illinois	55,410	2,539,891	30,128,405	2,456,578	129,921,395	42,780,851	10,944,790	822,441	5,249,274	465	5,739,249	36,083,405	2,747,339
Indiana	33,809	1,680,637	27,747,222	457,468	51,094,539	8,590,409	5,399,044	150,705	9,325,392	3	5,029,023	22,915,385	1,076,768
Iowa	55,045	1,194,020	29,435,692	505,807	68,935,065	21,005,142	5,914,620	84,209	71,792	....	2,967,043	27,512,179	1,777,339
Kansas	81,318	373,299	2,391,198	85,207	17,025,525	4,097,925	2,342,988	48,333	33,241	7	335,005	5,022,758	490,289
Kentucky	37,680	1,321,011	5,728,704	1,108,933	50,091,006	6,620,103	2,391,062	802,114	105,305,869	1,080	2,234,450	11,874,978	204,399
Louisiana	41,346	*857,039	9,906	934	7,596,628	17,782	67,695	1,083,575	15,541	350,832	140,428	822,405	8,776
Maine	35,000	626,915	278,706	84,115	1,089,888	2,351,354	7,771,009	854	15	....	1,774,168	11,636,482	1,053,415
Maryland	11,124	780,894	5,774,508	307,080	11,701,817	3,221,643	1,632,205	218,706	15,785,339	....	435,213	5,014,729	223,119
Massachusetts	7,800	1,457,351	34,644	229,227	1,397,807	797,664	3,025,446	917	7,312,885	....	306,659	8,589,161	597,455
Michigan	56,451	*1,344,031	16,265,773	144,508	14,086,238	8,954,466	10,318,799	3,651	5,385	....	8,726,145	24,400,185	1,290,923
Minnesota	83,531	446,056	18,866,073	78,088	4,743,117	10,678,261	1,943,063	1,064	8,247	....	401,185	9,522,010	695,053
Mississippi	47,156	827,922	274,479	14,858	15,637,316	414,586	214,189	1,743,432	61,012	564,938	288,285	2,613,521	8,324
Missouri	65,350	1,721,295	14,315,926	559,532	66,034,075	16,578,313	4,238,361	241,253	12,320,483	1,246	3,649,390	14,455,825	615,611
Nebraska	75,995	122,993	2,125,086	15,532	4,736,710	1,477,562	739,984	733	5,968	....	74,655	1,539,535	169,254
Nevada	104,125	58,711	228,196	319	9,063	55,916	199,249	....	95	106	27,029	110,880	31,855
New Hampshire	9,280	318,300	193,621	47,420	1,277,768	1,146,451	4,515,419	160	155,334	....	1,129,442	5,965,080	612,648
New Jersey	8,320	906,096	2,301,433	566,775	8,745,384	4,009,830	4,705,439	1,550,784	40,871	....	292,600	8,266,023	521,975
New York	47,000	*4,916,604	12,178,462	2,478,125	16,462,825	35,293,625	28,547,593	10,656	2,349,798	....	10,599,225	107,147,526	5,614,205
North Carolina	50,704	1,071,361	2,159,679	332,006	18,454,215	3,220,105	738,808	3,071,840	11,150,087	144,935	799,667	4,297,834	83,840
Ohio	39,964	2,665,260	27,882,159	846,890	67,501,144	25,347,549	11,192,814	220,295	18,741,973	....	20,539,643	50,266,372	2,289,565
Oregon	100,000	101,883	2,340,746	3,860	72,138	2,029,909	481,710	1,370	3,847	....	1,080,638	1,418,373	70,307
Pennsylvania	46,000	3,522,000	19,679,967	3,577,641	34,702,006	36,478,585	12,889,367	131,573	3,467,539	....	6,561,722	60,834,644	2,848,219
Rhode Island	1,306	*258,239	781	20,214	311,957	157,010	629,408	142	706	....	77,328	941,199	89,045
South Carolina	34,000	705,606	783,610	36,165	7,614,207	613,593	83,252	1,342,165	34,805	224,500	156,314	1,461,980	10,665
Tennessee	45,600	1,258,520	6,188,916	223,335	41,343,614	4,513,315	1,124,237	1,205,685	21,465,452	181,842	1,389,762	9,571,069	116,582
Texas	274,356	818,899	415,112	28,521	20,554,538	762,663	208,383	2,188,041	59,706	350,628	1,251,328	3,712,747	18,982
Vermont	10,212	330,551	454,703	73,346	1,699,882	3,602,430	5,157,428	95	72,671	....	3,102,137	17,844,396	1,020,669
Virginia	38,348	1,225,163	7,398,787	582,264	17,649,304	6,857,555	1,293,853	865,882	37,086,364	183	877,110	6,979,269	199,883
West Virginia	24,000	442,014	2,483,846	277,746	8,197,865	2,413,749	1,063,207	46,984	2,046,452	2	1,596,541	5,044,475	231,164
Wisconsin	53,924	*1,207,621	25,606,346	1,325,294	15,033,998	20,180,016	6,646,129	2,320	960,813	....	4,090,670	22,473,036	1,287,651
The Territories	1,619,350	780,000	1,845,164	17,663	1,189,385	1,098,837	963,655	664	11,568	29	1,204,590	1,854,768	90,098

For Physical and Commercial Chart of the World, see page 89.

Monteith's Comprehensive Geography.

DEFINITIONS.

1. **GEOGRAPHY** is a description of the Earth. It comprises Physical, Political, and Astronomical Geography.

2. **Physical Geography** describes the land and water divisions of the Earth, the climates, productions, and their effects upon mankind.

3. **Political or Civil Geography** describes the divisions which are formed for the purposes of government; as Republics, Kingdoms, States, etc.

4. **Astronomical Geography** refers to the Earth as one of the planets or bodies which revolve around the sun, and to the positions of places on the Earth's surface.

5. **The Earth was made** as a home for mankind. People live upon its surface.

If you stand on the shore and notice vessels coming in from sea, the sails of those most distant will appear to touch the water, the body being hidden by the curve of the Earth's surface.

6. **The shape** of the Earth is round, like a ball; we know this because navigators have sailed around it, and as ships come in from sea, their topmasts and sails appear first.

7. **The Earth is best represented** by a globe or sphere. One-half of a sphere is a hemisphere. The Earth's surface is usually represented by a map of the Eastern and Western Hemispheres, and sometimes by a map of the Northern and Southern Hemispheres.

8. **The Earth's surface consists** of land and water; one-fourth being land, and three-fourths water.

9. **The natural divisions of the land** are continents, islands, and peninsulas, which are diversified by mountains, plains, and valleys.

10. **Continents and Islands** are entirely surrounded by water.

11. **There are two Continents,** the Eastern and the Western.

12. **The Eastern Continent,** called the Old World, comprises Europe, Asia, and Africa; it is on the Eastern Hemisphere. The Western Continent, called the New World, comprises North and South America; it is on the Western Hemisphere. Each of these divisions is sometimes called a Continent, as the Continent of Europe. So, also, is Australia.

13. **There is more land** on the Eastern Hemisphere than on the Western; and more in the Northern Hemisphere than in the Southern.

14. **In the Center of all the land** are the British Isles; in the center of the Water Hemisphere is New Zealand. (See small map on page 87.)

15. **The largest division** of the Continents is Asia, and the smallest is Europe. North America is about twice as large as Europe, and Asia is nearly twice as large as North America.

16.

Grand Divisions.	Area in Square Miles.
Asia	16,400,000
Africa	11,550,000
North America	9,000,000
South America	6,900,000
Europe	3,800,000

17. **The shape** of the Grand Divisions is triangular.

18. **The division which has the longest coast line,** in proportion to its area, is Europe; Africa has the shortest.

Upon this depends much of the development of a country and the prosperity of the inhabitants.

19. **A Peninsula** is a portion of land almost surrounded by water. South America, Lower California, and Italy are Peninsulas.

20. **An Isthmus** (*ist'mus*) is a neck of land joining two larger portions of land. The Isthmus of Panama joins North and South America.

21. **A Cape** is a point of land extending into the water.

22. **A Promontory** is a high, rocky cape. Cape Horn is a Promontory.

23. **A Mountain** is a great elevation of land. A chain or range is an elevated ridge, or several mountains extending in a line. A mountain system is a number of chains in the same part of a continent. The Blue Ridge and Cumberland are ranges in the Appalachian or Alleghany system. The Rocky Mountain system includes the Rocky Mountain chain, the Sierra Nevadas, and the Sierra Madre.

24. **A Peak** is the top of a mountain, ending in a point.

Mt. Cervin, or the Matterhorn, Switzerland.—Point to the Mountains.—The Peak.—The Valley.—The Village.

25. **The highest peak** in a chain is called the culminating point.

26. **The most celebrated mountain systems** are the Rocky and Andes, on the Western Continent; and the Himalayas and Alps, on the Eastern Continent.

27. **The highest mountains** are about five miles high; but, as compared with the size of the Earth, they are like grains of sand on a globe ten inches in diameter. The highest mountain on the globe is Mt. Everest, in Asia, 29,000 feet high. The highest peaks in South America are four miles, and in North America and Europe three miles, above the level of the sea.

28. **Mountains exert an influence** upon climate, productions, and the pursuits of the inhabitants. Cold increases with the elevation, and perpetual snow covers the tops of very high mountains even in the hottest countries; on the sides of those mountains are all grades of temperature, and to the influence of their cold summits is the rain of some countries and states due.

Heat turns water into vapor, which rises; cold turns vapor into water in the form of rain and dew. Vapor frozen in flakes, is snow.

29. **Mountainous districts** are not adapted to agriculture, but great wealth is obtained from their mines; they contain extensive forests and pasture lands, and their streams afford water-power for mills and manufactories. Therefore the chief occupations of the inhabitants of mountainous districts are mining, manufacturing, grazing, and lumbering.

30. **A mountain pass** is a low part of a range, where travelers cross.

31. **A Volcano** is a mountain, whence issue fire, smoke, ashes, lava and steam. The opening is called a crater.

32. **The most celebrated volcanoes** are Vesuvius, in Italy; Etna, in Sicily; Hecla, in Iceland; and Cotopaxi (*ko-to-pax'-e*), in South America.

33. **Most of the volcanoes** on the globe are on the coasts and islands of the Pacific Ocean. The most remarkable volcanic region is in Malaysia.

34. **A valley** is land between hills or mountains. A vale is a small valley.

35. **Valleys are low** as compared with the hills or mountains which surround them. In mountainous districts, the valleys are considerably elevated above the sea level. Those of the Caspian, Ar'al and Dead Seas are below the level of the sea.

36. **The soil of valleys** is usually fertile.

37. **A Plain** is a level tract of land. The grassy and treeless plains of North America are called prairies or savannahs; of South America, llanos (*lyah'-noce*) and pampas; of Russia, steppes (*steps*).

38. **The forest or wooded plains** of South America are called silvas. A swamp, marsh, moor, or fen, is land which is usually or occasionally covered with water.

39. **Plains cover** about two-thirds of the surface of the Western Continent.

Note.—The paragraphs in fine type need not be committed to memory; they are designed to suggest remarks and explanations.

Mt. Hecla, Iceland.—Point to the Volcano.

The Coast of Asia Minor.—Point to the Bay.—The Cape.—The Coast.—The Strait.

49. **The movements** of the water of the ocean are three: waves, tides, and currents.

Waves are caused by winds and tides; tides, by the influence of the moon and sun; and oceanic currents, by the combined action of the wind, the daily revolution of the earth, and differences in the temperature of the water.

50. **The currents** cause a constant interchange of the waters between the hot and the cold regions.

51. **The warm currents** are the Equatorial Current, the Japan Current, and the Gulf Stream; the Arctic and Antarctic Currents are cold.

52. Without the warming influence of the Gulf Stream, Great Britain and Ireland would be as cold and unproductive as Labrador and Newfoundland.

53. The warm current of the Pacific Ocean washes the western coast of North America, and gives to California and Oregon a much milder climate than that of the Atlantic States in the same latitude. East and west of Greenland, cold currents flow southerly, and bring icebergs as far as Newfoundland, where they are melted by the warm water of the Gulf Stream.

40. **Plateaus** are elevated tracts of land comparatively level. The highest plateaus are in Asia. Their elevation is about 15,000 feet. The Himalaya mountains rest upon them.

Nearly the whole of Mexico is a plateau, which has an elevation of more than 7,000 feet. The City of Mexico is 7,400 feet above the level of the sea.

41. **A Desert** is a barren region of country; its barrenness proceeds, mainly, from lack of rain. Some deserts are sandy, as in Africa; some are salt, as in Persia; and others are stony, as in Arabia.

42. **An Oasis** is a fertile spot in a desert, where trees and grass abound; its fertility is due to springs.

43. **An Ocean** is the largest division of water; there are five oceans—the Pacific, Atlantic, Indian, Arctic, and Antarctic. The oceans, considered together, are called the sea.

44. **The large arms** of the ocean are called seas, gulfs, and bays.

45. **The largest ocean** is the Pacific; but the most important to America and Europe is the Atlantic.

46. **The principal arms of the Atlantic Ocean** are the Mediterranean, Baltic, North and Caribbe'an Seas, the Gulfs of Mexico and St. Lawrence, the Bay of Biscay, and the English Channel.

The area of the Pacific is greater than that of all the land on the globe; its islands are numerous, and its inlets are mainly on the coast of Asia. Its inlets on the Pacific coast of America are few, owing to the high mountains near it.

47. **The water** of the sea is salt. Fresh water is that which has been raised from the ocean by evaporation and returned to the land in the form of rain or snow.

48. **The depth** of the sea is greatest in the Torrid Zone—about six miles.

The Atlantic cables between Ireland and Newfoundland lie on the bottom, at a depth of two miles.

The ocean is essential to the life of mankind; its vapors supply rain, springs, and streams; it tempers the climate, and furnishes easy communication between nations.

54. **A Strait** is a passage of water connecting two large bodies of water.

The Strait of Gibraltar connects the Mediterranean Sea with the Atlantic Ocean.

55. **A Sound** is a shallow channel or bay.

56. **A River** is a stream of water flowing through the land. Its head is where it rises; its mouth is where it flows into another river or body of water.

57. **A River is formed** by springs, or is the outlet of a lake. A spring is rain or snow-water coming out of the ground.

58. **Rivers are useful** as a means of communication between different parts of a country. From them cities and towns are supplied with fresh water, and mills and manufactories with water-power.

The most important river in North America is the Mississippi. The water of some rivers, like the Nile, flowing through dry countries, is used for watering and fertilizing the soil.

The Plain of Languedoc, France.—The Pyrenees.—Point to the Plain.—The Mountains.

The Waterfall of Staubbach, Switzerland. A view in the Valley of Lauterbrunnen, looking south. In the distance is the high mountain called the Jungfrau, which is covered with perpetual snow for a considerable distance below its summit. A boy blowing an Alpine Horn.

59. **The right bank** of a river is on your right hand as you descend the river. Its upper course begins at its source, and passes over that portion of its bed which is most inclined; this part usually contains rapids and waterfalls.

Some rivers, like the Nile and the Mississippi, empty through several mouths; the land enclosed by those mouths is called a delta.

60. **Up a river** is toward its source; down a river, toward its mouth, or with the stream.

61. **The Basin** of a river comprises all the land drained by it and its branches.

The basin of the Mississippi covers an area of 1,000,000 square miles; that of the Amazon is twice as large.

62. **A Watershed** is the ridge of land which surrounds a river basin, or the place where waters separate.

63. **The Course** or direction of a river is always governed by the slopes of the land over which it flows.

64. **The deepest part** of a stream is called the channel.

65. **A Cascade, Waterfall, or Cataract,** is a stream dashing down a precipice.

Niagara Falls are celebrated for their mass of water; the Yosemite Falls, in California, for their great height—2,600 feet. The waterfalls of Europe are numerous and picturesque; those of Staubbach, in Switzerland, are 800 feet high.

66. **Rapids** are water rushing down an inclined bed.

67. **A Confluence** is where two or more rivers meet. Affluents and tributaries of a river are the streams which flow into it.

68. **A Canon** (*kan'yon*) is a gorge between high steep banks, worn by a stream.

Mont Blanc; Valley of Chamouni; a glacier extending from the region of perpetual snow down to the valley.
The River Arve, which empties into the Rhone near Geneva.

69. **An Estuary, Frith, or Firth,** is a narrow and deep inlet of the sea, at the mouth of a river.

70. **A Glacier** (*gla'seer*) is a great mass of snow and ice which moves slowly down the sides of a mountain, bearing with it sand and fragments of rocks.

From the lower end of a glacier a stream issues, which is the beginning of a river.

71. **An Avalanche** is a large mass of snow, ice, and earth, sliding or rolling down the side of a mountain.

The Alps are celebrated for avalanches and glaciers.

72. **A Lake** is a body of water almost surrounded by land.

Lakes are supplied from rain, springs, and streams; in mountain regions, from melted snow. The water of most lakes is fresh. Salt lakes are those which have no outlets. Lakes are at various elevations; Lake Titicaca, in South America, is two miles above, while the Dead Sea is 1300 feet below, the level of the sea.

73. **The largest body of fresh water** on the globe is Lake Superior; its area is about three-fifths that of England.

	Area in Square Miles.
Lake Superior	32,000
Lake Erie	9,000
Lake Champlain	600
Lake Geneva, Switzerland	80
Loch Lomond (largest in Great Britain) . .	40
Lake Windermere (largest in England) . .	10

Lakes in Scotland are called lochs; in Ireland, loughs.

74. **The most celebrated Salt Lakes** are the Caspian, Ar'al, and Dead Seas, and Great Salt Lake.

ASTRONOMICAL GEOGRAPHY.

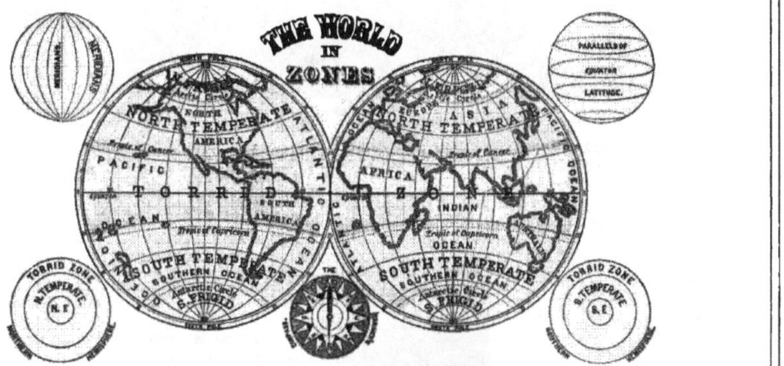

1. **The Earth** is a sphere or globe; its diameter is the distance through it, or a line passing from any point on its surface through the center to the opposite point.

The diameter of the Earth is about 8,000 miles, and its circumference, or distance around it, is about 25,000 miles.

2. **It moves** rapidly and constantly around the sun, from which it receives light and heat.

It requires 365¼ days for the Earth to make one revolution around the sun, moving at the rate of over a million miles every day. If the Earth did not revolve around the sun, we would have no change of seasons.

3. **The light** of the sun covers one-half the Earth's surface at a time; on that side it is day; on the opposite side, night.

The Succession of Day and Night. The lamp represents the Sun: the apple, the Earth; the needle on which the apple turns represents the Axis of the Earth.

4. **The succession of day and night** is caused by the revolution of the Earth on its axis.

5. **The axis of the Earth** is an imaginary line, on which it performs its daily motion.

6. **The two points where it touches the surface** are called the North and the South Pole.

7. **Lines drawn on the Earth's surface** from Pole to Pole are called Meridians, which always extend north and south; the circles which pass from east to west are the Equator and parallels of latitude. The Meridians are perpendicular to the Equator.

8. **There are two kinds of Circles,** great and small.

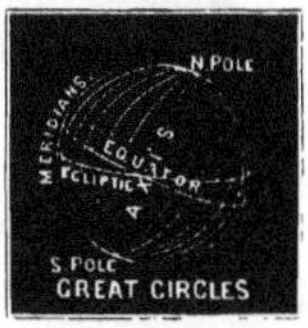

9. A Great Circle divides the Earth into two equal parts, while a Small Circle divides it into two unequal parts.

10. **The Equator** is a great circle which divides the Earth into Northern and Southern Hemispheres.

The points, lines, and circles, on artificial globes, are not really on the surface of the Earth; they are only imagined to be there, for the purposes of measuring distances and determining the positions of places on the globe.

11. **A Degree** is a three hundred and sixtieth part of a circle. It varies in length, according to the size of the circle. On a great circle it is equal to 69¼ statute miles. Degrees are represented by (°).

12. **The principal small circles** are the two Tropics and two Polar Circles.

13. **The Tropics** are parallel with the Equator, and about 23½ degrees distant from it. They are the Tropic of Cancer and the Tropic of Capricorn.

14. **The Polar Circles** surround the Poles, and are 23½ degrees distant from them. The North Polar or Arctic Circle surrounds the North Pole, and the South Polar or Antarctic Circle surrounds the South Pole.

15. **Latitude** is distance north or south from the Equator.

Places on the Equator have no latitude; the North and South Poles are in the greatest latitude— 90 degrees, or a quarter of a circle (*Hemi* or *Semi*-half).

16. **Longitude** is distance east or west from a certain meridian.

We measure longitude from the meridian of Greenwich and the meridian of Washington.

The places having the greatest longitude are under the meridian which is opposite that from which longitude is reckoned; their longitude is 180°.

When the North Pole leans toward the Sun, the Northern Hemisphere has summer, long days, and short nights; the Sun is vertical to the inhabitants at the Tropic of Cancer.

When the North Pole leans away from the Sun, the Northern Hemisphere has winter, short days, and long nights; the Sun is vertical to the inhabitants at the Tropic of Capricorn.

17. **Latitude is marked** on the right and left sides of maps; longitude, on the upper and lower sides. On maps of the Hemispheres, longitude is marked on the Equator.

Turn to p. 16. What is the latitude of Philadelphia? Of Cape Farewell? Of the southern part of Cuba? What is the longitude of Washington from the meridian of Greenwich?

18. **Zones** are the five great belts into which the Earth's surface is divided by the Tropics and the Polar Circles. There are five zones: the Torrid, North and South Temperate, and North and South Frigid.

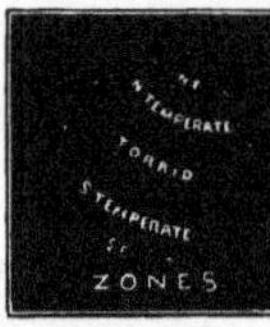

19. **The Torrid Zone** lies between the Tropics. It is 47 degrees (47°) from north to south, and the Equator is in the middle of it.

20. **The North Frigid Zone** is between the North Pole and the Arctic Circle. In its center is the North Pole.

21. **The South Frigid Zone** is between the South Pole and the Antarctic circle.

22. **The North Temperate Zone** is between the Torrid and the North Frigid Zone.

23. **The South Temperate Zone** is between the Torrid and the South Frigid Zone; the Temperate Zones are each 43° wide.

24. **The Torrid Zone** is hot throughout the year, because the sun shines more directly on that part of the Earth's surface. To every place in the Torrid Zone the sun is vertical (directly overhead) at certain times in the year.

The sun is never vertical to the inhabitants of the Frigid or Temperate Zones. Vertical means directly overhead, or in the zenith.

25. **The Frigid Zones** are cold throughout the year, because the sun shines indirectly or obliquely on those parts of the Earth's surface.

26. **Within the Temperate Zones** the heat is less than that of the Torrid, and the cold less than that of the Frigid Zones. Here are enjoyed four seasons—Spring, Summer, Autumn, and Winter.

27. **The change of Seasons** is caused by the revolution of the Earth around the Sun, and a uniform inclination of the Earth's axis to the plane of its orbit. (See paragraph 2).

28. **The Earth's Orbit** is the path or curved line in which it revolves around the Sun.

29. When the North Pole leans toward the Sun, it is summer in the Northern Hemisphere; six months afterward, the Earth will be on the opposite side of the Sun, when the North Pole will lean away from the Sun, and the Northern Hemisphere will have winter.

The seasons in the Southern Hemisphere are always the reverse of those in the Northern.

The Tropics mark the limits within which the inhabitants may have a vertical Sun. Tropic means turning.

The Polar Circles mark the limits within which the days and nights can be more than 24 hours long.

On the 21st of June, the Sun is above the horizon to places on the Arctic Circle, during the whole revolution of the Earth on its axis; that is, their daylight continues 24 hours. As you leave that circle and approach the North Pole, the length of the day increases, until you reach the North Pole, where daylight continues for six months, the Sun rising in March and setting in September.

30. **The Sun is vertical** to the inhabitants at the Equator in the latter part of March and September, when the days and nights are equal throughout the world.

31. **The Sun is vertical** to the inhabitants at the Tropic of Cancer in the latter part of June, when the Northern Hemisphere has summer, long days and short nights.

32. **The Sun is vertical** to the inhabitants at the Tropic of Capricorn in the latter part of December, when the Southern Hemisphere has summer, long days and short nights.

33. **The Sensible Horizon** is the circle which bounds our view of the Earth's surface; it is best seen on the ocean, or on a plain, where the view is not obstructed by houses, hills, etc.

34. **The Cardinal Points of the Horizon** are North, East, South, and West.

If you face the north, the east will be on your right hand; the west, on your left; and the south will be behind you.

The north is shown by a mariner's compass, a box containing a needle which always points in that direction. (See illustration on page 9.)

35. **The Ecliptic**, in geography, is a great circle on the globe which is always in the plane of the Earth's orbit.

BLACKBOARD EXERCISE.

Draw a circle and give a definition of it. Supposing that circle to represent the Earth, draw its axis, as shown in the picture above paragraph 9. What is the Earth's axis?

Mark the North and the South Pole. What are they?

Draw the Equator. What is it? Draw the Ecliptic. What is it? Draw Meridians. Between what two points are they drawn? With what Great Circle do Meridians form right angles?

On this or another circle mark the small circles. Then the Zones, as shown opposite paragraph 18. Between what circles is the Torrid Zone? The North Temperate? The South Temperate Zone? A line drawn through the center of the Sun and Earth will represent the plane of the Ecliptic or the Earth's orbit. Draw a circle to represent the Earth, then the axis on the 21st of June, as shown at head of the page. Which Pole leans toward the Sun at that time? Draw an upright line, as here shown, to separate day from night. The top and bottom of this line fix the Arctic and Antarctic Circles; and the two points where the Ecliptic here appears to intersect the surface fix the Tropics.

The teacher can easily explain from the illustration above the cause of long days and nights at the Poles; as the pupils will readily see that every revolution of the Earth on its axis does not bring day and night to places that are near the Poles, as it does to other parts of the Earth's surface.

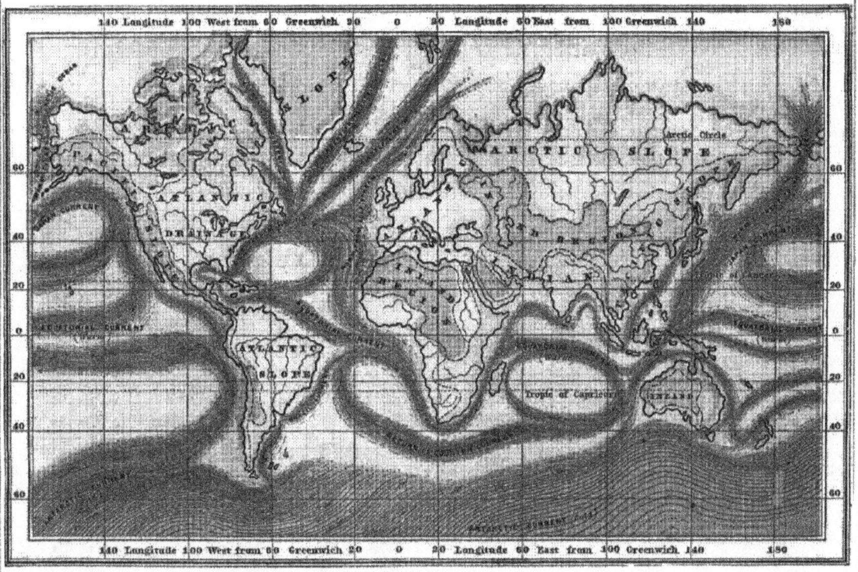

Oceanic Currents.

CLIMATES AND THEIR EFFECTS.

1. **Climate** is the condition of a place or country in relation to the temperature and moisture of its atmosphere.

2. **It depends** upon the latitude, elevation, winds, oceanic currents, and mountain ranges.

3. **Moisture** is vapor which rises from the ocean and other bodies of water on the Earth's surface. It is carried over the land by the wind, and when it enters cold air, it becomes rain or snow.

4. **More rain falls** on the coasts than in the interior of a continent; and more on that side of a continent or mountain chain against which the prevailing winds blow.

5. **The Zone** in which the most rain falls is the Torrid.

6. **On the Western Continent,** the greatest amount of rain falls in South America, between the Andes Mountains and the eastern coast, where the prevailing winds are from the east.

7. **In the Temperate Zones,** the winds blow from the west or southwest; therefore much rain falls on the western coasts of North America and Europe.

8. **The great rainless regions** are in the interior of Africa and Asia. A region without moisture is a desert.

9. **The heat diminishes** as you leave the Torrid Zone and travel toward either Pole, or as you ascend a high mountain.

From the Equator toward the North Pole, the temperature diminishes about 1° for every 100 miles; and from the level of the sea to the summit of a mountain, the temperature diminishes about 1° for every 350 feet.

10. **Coasts** that are washed by warm oceanic currents have a warmer climate than other parts of a continent in the same latitude.

11. **The western coast of Europe** is washed by the Gulf Stream, a warm current which, with the aid of the westerly or southwesterly winds, gives to that part of Europe a climate celebrated for its mildness and moisture, while that of the eastern part is very cold in winter and very hot in summer.

12. **The western coast of the United States** is washed by a warm current of the Pacific Ocean, giving to California, Oregon, and Washington Territory, a climate similar to that of Western Europe.

The land is warmer in summer and colder in winter than the ocean; consequently, winds which blow over the ocean are more even in temperature than those which blow over the land.

13. **Climates produce important effects** upon the vegetation of different countries and upon the condition and pursuits of the inhabitants.

14. **Vegetation is most luxuriant** in the Torrid Zone; this is due to the great heat and moisture of that region.

MANKIND.

The Caucasian Race: Egyptian, Arab, Abyssinian, European.

1. **Mankind is divided** into five general classes or races: the Caucasian, or white race; the Mongolian, or yellow race; the Malay, or brown race; the American Indian, or red race; and the Ethiopian, or black race.

2. **The races are distinguished** from each other by the form of the head and face, the kind of hair, and the color of the skin.

3. **In the Caucasian race** the head is almost round, the nose narrow and prominent, the mouth small, and the **hair** long. Although the skin is mostly white, or of a light shade, yet some Caucasians are quite dark.

4. **The Caucasians include** most Europeans and their descendants, besides the inhabitants of Western Asia and Northern Africa. The Egyptians, Moors, Berbers and Ar'abs are of a dark color, and the Abyssinians are black; they nevertheless belong to this race.

5. **The Caucasian** is the most enterprising and enlightened race, especially the inhabitants of the North Temperate Zone.

Mongolian or Yellow Race: Esquimaux and Chinese.

6. **The Mongolians** include the Chinese, Japanese and Esquimaux. They are short in stature, and have broad faces, low foreheads; wide, small noses; coarse, straight hair.

7. **The Malays** inhabit the Malay Peninsula, Sumatra, Java, New Zealand, the Sandwich Islands, and many other islands of the Pacific and Indian Oceans.

8. **Their color** is reddish-brown, and their hair is black, coarse, and abundant.

9. **Their character** is savage and treacherous.

The Malay Race: Sandwich Islanders and New Zealanders.

10. **The American Indians** are copper-colored, and tall in stature. They have prominent cheek bones, and long, straight hair.

American Indians, or Red Race.

11. **The Black Race** includes the inhabitants of nearly all that part of Africa which is south of the Great Desert, besides large numbers in North and South America and the West Indies. The nose of this race is broad; lips, thick; and hair, woolly.

The Black Race: Caffres, and Natives of Western Africa

12. **People differ** in their conditions and occupations.

13. **Savages live** by hunting and fishing. Half civilized tribes own cattle, horses, and sheep, and move their tents from place to place to find pasture. The chief occupations of civilized nations are agriculture, mining, manufacturing, and commerce.

14. **Agriculture** is conducted by the farmer; mining, by the miner; manufacturing, by the manufacturer; and commerce or trade, by the merchant.

15. **Mining** is digging for metals or other minerals, as gold, silver, lead, iron, coal, and salt. Quarrying is taking out building stone, as granite, marble, etc.

16. **Commerce** is the exchange of products between different countries or states. The goods or products sent from a country are exports; those brought into it, imports. Foreign commerce is conducted between different countries; domestic commerce, between different parts of the same country. Products are conveyed on the sea, rivers, canals, and railroads.

17. **The Political Divisions** of the world include Republics, Empires, Kingdoms, States, etc.

18. **A Republic** is a country or nation whose laws are made and executed by men elected by the people; as the United States, Peru, and Switzerland.

19. **An Empire** is a region comprising several countries governed by an Emperor; as Russia, Germany, and the Chinese Empire.

20. **A Kingdom** is a country governed by a King or a Queen; as Italy, Spain, and Denmark.

21. **A Monarchy** is a government in which the supreme power belongs to one person, called a monarch, who inherits the office.

22. **A Limited or Constitutional Monarchy** is a government in which the power of the monarch is limited by law; as Great Britain, Prussia, and Brazil.

23. **In an Absolute Monarchy or Despotism** the power of the ruler is unlimited; as in Russia and the Chinese Empire.

24. **The governments** of the Old World are mostly monarchies; of the New World, republics.

The Emperor of Russia is called the Czar; of Turkey, the Sultan or Caliph; the King of Egypt, who is subject to the Sultan, is called the Khedive. *Kaiser*, in Germany, Emperor.

25. **Federal Republics are composed** of states which are independent in the management of their local affairs, but united under one general government.

26. **States are divided** into counties, which contain cities, towns, and villages.

27. **A Village** is a small collection of houses and inhabitants; towns are larger than villages; cities are large towns having special privileges.

28. **The Capital** of a state or country is the city in which the laws are made, and where the chief officer resides.

29. **The Metropolis**, or chief city, is that which contains the largest number of inhabitants.

30. A city, town, or village, is generally located with reference to some natural features, such as, on a bay or harbor, where ships may anchor safely; on a navigable river; at the junction of two rivers; where water-power can be obtained for mills and manufactories; near mines or quarries; or at the end of a lake.

HISTORICAL GEOGRAPHY.

1. **Adam and Eve** were placed in the Garden of Eden (B. C. 4004). Eden was probably situated in the western part of Asia.

2. **About 2000 years after,** their descendants were destroyed by the flood, except only Noah and his sons and their wives.

3. **Some time after the flood,** Noah's descendants were scattered; those of his son Shem settled in Asia; of Ham, in Africa; and of Japheth, in Europe.

4. **The first inhabitants of America** were, probably, adventurers from Asia, across Behring Strait.

5. **The ancient Egyptians** were celebrated for their civilization and learning; and the Phenicians, for their skill in navigation and commerce.

6. **The Phenicians,** or Canaanites, inhabited the eastern coast of the Mediterranean, and explored all the coast of that sea, besides the western shores of Europe and Africa (800 B. C.).

7. **The great monarchies of ancient times** were Assyria and Babylon, Persia, Greece, and Rome.

8. **Rome was most powerful** about the beginning of the Christian era; among its most celebrated rulers were Julius Cæsar and Augustus Cæsar.

9. **The Middle or Dark Ages** were from the fifth to the fifteenth century. Identified with them were the rise and progress of Mohammedanism, the Feudal System, and the Crusades.

10. The Turks having conquered Syria, were very cruel to the Christian pilgrims who visited Palestine; consequently, many thousand Christians throughout Europe left their homes for that distant land, to drive out the Turks. Although great numbers perished on the way, the Christians were successful.

These expeditions, which occurred in the 11th, 12th, and 13th centuries, were called the Crusades. The Turks again obtained possession of Palestine, and have held it ever since.

11. **The 15th century was remarkable** for important discoveries; among them was that of America, by Columbus.

12. **In the 16th century** the English and Dutch made efforts to reach India by a northwest passage around the northern part of America.

13. **Among the celebrated explorers** of that route were Frobisher, Davis, Hudson, and Baffin.

14. **The first voyage around the globe** was made by Magellan, in the 16th century; and another by Captain Cook, in the 18th century, who made known the great length of the Arctic coast of North America.

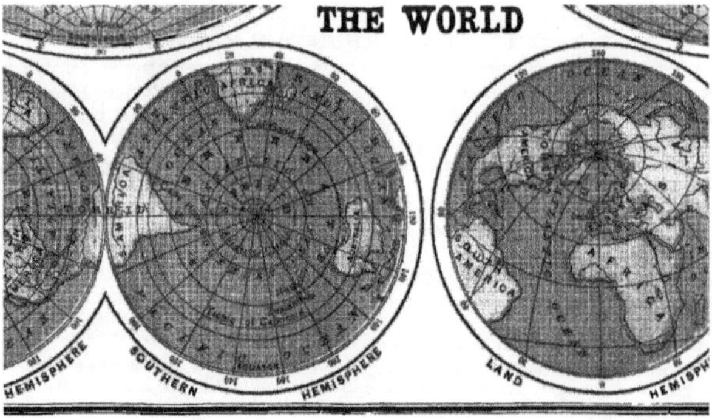
THE WORLD
HEMISPHERE
SOUTHERN
HEMISPHERE
LAND
HEMISPH
AFRICA
SOUTH AMERICA
AFRICA

THE HEMISPHERES.

LESSON I.

1. **The Eastern Hemisphere** contains Europe, Asia, Africa, and Australia.

2. **The Western Hemisphere contains** North and South America; a small portion of Asia is in this Hemisphere.

3. **The Northern Hemisphere** is that half of the Earth which is north of the Equator; the North Pole is its center.

4. **It contains** North America, Europe, Asia, the greater part of Africa, and the northern part of South America.

5. **The Southern Hemisphere** is that half of the Earth which is south of the Equator; the South Pole is its center.

6. **It contains** the greater part of South America, the southern part of Africa, and the whole of Australia.

7. **The Land Hemisphere** contains all the continents except the southern part of South America; Europe is in its center.

8. **The Water Hemisphere** contains Australia and the southern part of South America; New Zealand is its center.

9. **The longest straight line** that can be drawn on the land surface of the Earth is from the western part of Africa to the northeastern part of Asia, about 11,000 miles.

10. **The highest Mountains** and table-lands are in Asia and South America.

GENERAL QUESTIONS.

LESSON II.

Which Hemisphere contains the more land, the Eastern or the Western? The Northern or the Southern?

Which division of land extends furthest north? Which extends furthest south? What three divisions are wholly in the Northern Hemisphere? Which are partly in the Northern and partly in the Southern?

Is the greater part of South America in the Northern or the Southern Hemisphere? Is the greater part of Africa in the Northern or the Southern Hemisphere? What large island is in the Southern Hemisphere? What two large Islands are crossed by the Equator?

In what Zone is the greater part of South America? Africa? North America? Europe? Asia? What part of North America is in the Torrid Zone? What part of Asia? Is any part of Europe in the Torrid Zone? In what three Zones is North America? Asia? Africa? In what two Zones is South America? Europe? Australia?

Which is the largest of the Grand Divisions? The smallest?

LESSON III.

What two straits in the Western Hemisphere are crossed by the Arctic Circle? What gulf in the Western and two seas in the Eastern Hemisphere are crossed by the Tropic of Cancer? What two large islands in the Eastern Hemisphere are crossed by the Tropic of Capricorn?

What Mountains in North America? In South America? In Africa? In Asia? Between Europe and Asia?

What Rivers in North America? In South America? In Africa? In Asia?

What Gulfs and Bays in North America?

What Sea north of South America? Between Europe and Africa? What two between Europe and Asia? South of Asia? What sea northwest of Europe?

What Cape on the northern coast of South America? On the eastern? On the southern? On the western? On the southern coast of Africa? On the western coast of Africa? On the northern coast of Europe? On the southern coast of Asia?

REVIEW QUESTIONS.

LESSON IV.

Where are they?

The Grand Divisions:—North America, South America, Europe, Asia. Africa.

Islands:—Newfoundland, The British Isles, Greenland, The West Indies, The Japan Islands, Madagascar, New Zealand, Australia, Borneo.

Mountains:— Atlas, Rocky, Andes, Himalaya, Ural, Snow.

Rivers:—Mississippi,* Amazon, Nile, Parana, Cambodia. Yang-tse Kiang, Obi, Niger.

Seas, Gulfs, and Bays:—Caribbe'an Sea, Caspian Sea, Mediterranean Sea, Hudson Bay, Baffin Bay, Black Sea, China Sea, Arabian Sea Gulf of Mexico, North Sea, Gulf of Guinea.

Capes:—Horn, Farewell, Geod Hope, Verd, St. Roque, North.

* *In describing rivers* throughout the book, tell where they rise, in what directions they flow, and into what they empty.

SEA VOYAGES.

Remember that all meridians run north and south, and all parallels of latitude, east and west.

Observe that Greenland points to the south, not southeast, and that Cape Farewell is nearly due north of Cape St. Roque.

A globe should be here shown; also chart on page 89.

On what waters and in what directions would you sail from New York to Cuba? Cape St. Roque? Newfoundland? Baffin Bay? Rio Janeiro? British Islands? Mediterranean Sea? Cape of Good Hope? Cape Horn?

What is the shortest route by water from New York to Asia?

What directions would you take, and what capes would you pass, in a voyage from New York to Australia? To New Zealand? To Iceland? To Nova Zembla? To the Japan Islands? To San Francisco? What is the shortest route from New York to San Francisco? *Ans. By railroad.* From New York to the Sandwich Islands? From New York to Japan?

If you would sail westwardly from the Sandwich Islands, or on the parallel of 20 north latitude, at what part of Asia would you arrive?

If you should sail westwardly from San Francisco, at what islands on the Eastern Hemisphere would you arrive?

If you should sail eastwardly from New York, at what part of the Eastern Continent would you arrive?

By what two routes can you sail from San Francisco to Japan? New York to Australia? New York to Japan?

What islands and capes would you pass in a voyage by water around the world, starting from New York? Starting from San Francisco?

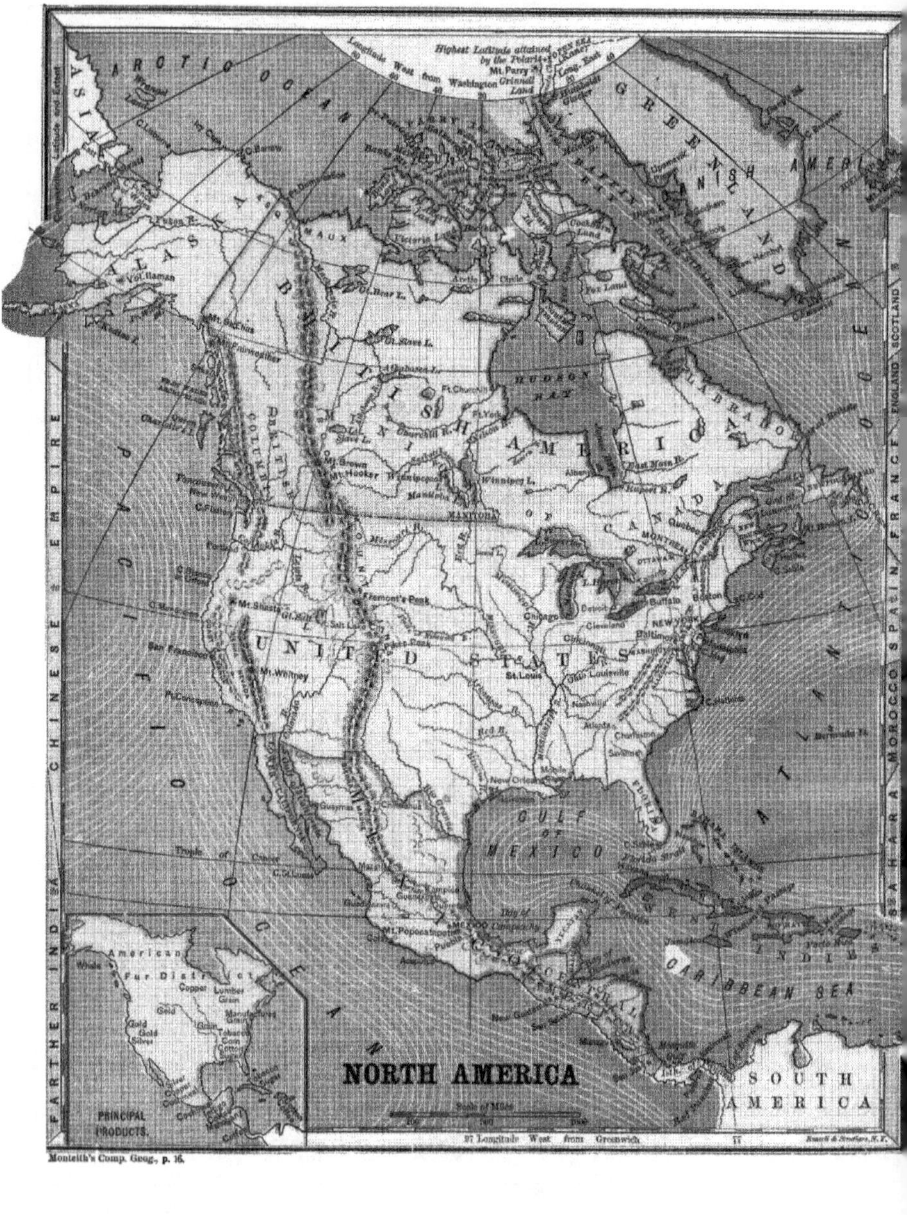

Monteith's Comp. Geog., p. 16.

LESSON V.

NORTH AMERICA.

1. **North America is situated** in the Western Hemisphere, and lies chiefly in the North Temperate Zone.

2. **Its most northern** part is in the North Frigid Zone, and its most southern, in the Torrid Zone.

3. **Its widest part** is in the north.

4. **Its direction** from South America is northwest; from Europe, west; and from Asia, east.

5. **Its shape** is that of a triangle, and its three sides are bounded by three oceans—the Arctic, Atlantic, and Pacific.

6. **Its size** is twice that of Europe, or one-half that of Asia.

GENERAL QUESTIONS.

What countries are crossed by the Arctic Circle? What country and gulf are crossed by the Tropic of Cancer?

Through what country does the parallel of 40° north latitude pass? Through what large bay does the parallel of 60° north latitude pass? On which side of North America are its long mountain ranges? Name those ranges. On which side of North America are its large gulfs and bays? Name them.

What capes project into the Arctic Ocean? Into the Atlantic Ocean? Into the Pacific Ocean?

Name the countries of North America. What division of land is Greenland? *Ans.* An island. To what government does Greenland belong? To what government does British America belong? What large rivers are west of the Rocky Mountains? What large rivers flow into the Gulf of Mexico? Into Hudson Bay? What river flows into the Arctic Ocean?

Bound British America. The United States. Mexico. Central America.

What countries on the Eastern Continent are in the same latitude as British America? (See right and left margins.) As the United States? What meridian passes through the centre of North America and near the western coasts of Hudson Bay and the Gulf of Mexico?

General Directions for Drawing the Continents

Make a scale by marking on a slip of paper or pasteboard the measures *1*, *2*, *3*, etc., as on the sides of this map. Every measure on the maps of the continents represents *six hundred miles*.

All names in black type are to be marked on the drawing.

Pupils should mark on each river, ocean, bay, gulf, strait, etc., the first syllable of its name, when there is not room enough for the full word.

As an exercise in spelling, the full name of each abbreviated word may be written on the margin of the drawing, before the recitation closes.

NOTE.—*Do not draw the measurement lines.* The marking of the *points* indicated will be sufficient.

SEA VOYAGES.

What capes would you pass on a voyage from Philadelphia to Savannah? What islands would you pass on your way from New York to Cuba? From Boston to Quebec?

On what waters would you sail from Savannah to New Orleans? From Quebec to Hayti? From New York to Baffin Bay?

LESSON VI.

TO DRAW NORTH AMERICA.

Begin at **A**: *1* measure north of **A**, draw ***Hayti***; *5* ms. north, draw ***Newfoundland*** and the ***Strait of Bellisle.*** Between *6* and *7* draw the eastern coast of ***Greenland.***

From **C** measure to **D**, marking the points *1*, *2*, *3*, *4*, *5*. Through *1* draw the northern extremity of the ***Peninsula of Yucatan.*** *1* m. *west of this point* draw the western coast of the ***Gulf of Mexico.*** Draw ***Campeachy Bay***, the ***Bay of Honduras***, and the ***Isthmus of Panama***, and ***Cuba.***

At *2* draw the ***Peninsula of Florida.*** Mark *3* on the line **C D**, and locate ***Cape Hatteras.*** Near *4*, mark ***Cape Cod;*** and near 5 draw the ***Gulf of St. Lawrence*** and ***Nova Scotia***, and complete the eastern coast. From **B** measure *5* ms. to **H**, and draw the islands along the Arctic coast.

From **G**, measure north *3* ms., and mark ***San Francisco;*** and a little to the north, ***Cape Mendocino***, the western cape of California; opposite *4*, draw ***Vancouver's Island;*** at *5*, ***Queen Charlotte's Island.*** *7* is near the northeastern coast of ***Alaska.***

From **E** toward **F**, mark the points *1*, *2*, and *3*, and draw the ***Gulf*** and ***Peninsula*** of ***California*** and ***Cape St. Lucas.*** Complete the southern coast of the continent.

Draw ***James Bay, Hudson Bay, Great Slave Lake***, and ***Mackenzie River***, according to the line **I J**, and complete the northern part of the continent.

When the boundaries are drawn, add the mountains, rivers, countries, bays and gulfs, capes, islands, and cities; then the section from Cape Hatteras to the Pacific Ocean, showing elevations above the level of the sea.

REVIEW QUESTIONS.

LESSON VII.

Where are they?

Mountains:—Rocky, Alleghany, Sierra Nevada, St. Elias.

Rivers:—Mississippi, Rio Grande, Mackenzie, Athabasca, St. Lawrence, Columbia, Missouri, Colorado, Yukon.

Gulfs and Bays:—Gulf of St. Lawrence, Bay of Campeachy, Bay of Honduras, Gulf of Mexico, Hudson Bay, Fox Channel, Baffin Bay, James Bay.

Straits:—Hudson, Davis, Behring, Florida.

LESSON VIII.

Where are they?

Lakes:—Superior, Great Slave, Michigan, Great Bear, Winnipeg, Ontario, Erie, Hu'ron.

Islands:—New'foundland, Vancouver's, Hayti (*hay'te*), Queen Charlotte's, Jamaica, Anticosti, Bermuda Islands, Iceland, Cuba.

Capes:—Farewell, Sable (two capes), Hatteras, Flattery, Race, Mendocino, St. Lucas.

The pupils may also point out these places on the Relief Map below.

PHYSICAL AND DESCRIPTIVE GEOGRAPHY.

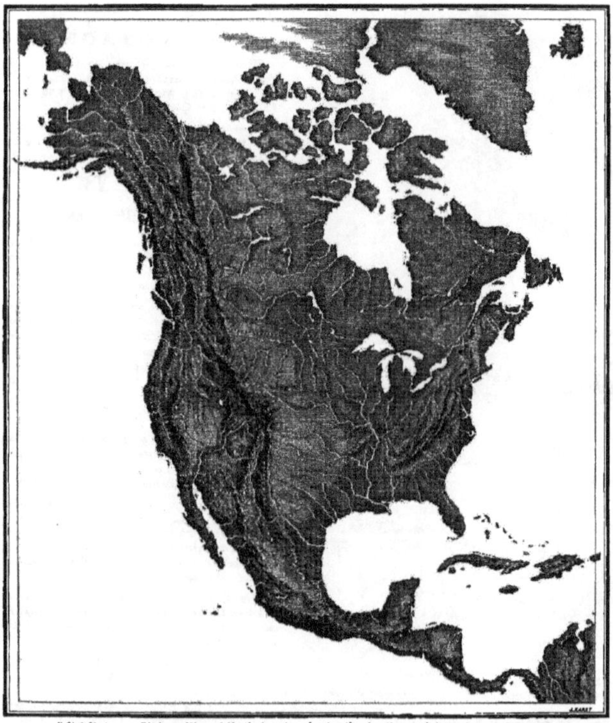

Relief Map, or a Birds-eye View of North America, showing the elevations and depressions of the surface.

LESSON IX.

1. **The mountainous section** of North America is in the west.

2. **Its great mountain system** extends from the Andes, in South America, in a northwesterly direction, to the Arctic Ocean.

3. **The mountain ranges** are mainly parallel with the sea-coasts.

4. **The Rocky Mountains** rest on a great plateau which extends westerly to the Sierra Nevada and Cascade Mountains. Their greatest distance from the Pacific coast is about 1000 miles. In the United States they are about midway between the Pacific Ocean and the Mississippi River. The plateau is about one mile above the level of the sea.

5. **The highest peaks** of the Rocky, Sierra Nevada and Cascade Ranges are so high that they are continually covered with snow. (See page 6, paragraph 28.)

6. **This section is celebrated** for its wealth in gold, silver, and quicksilver mines. In the valleys near the Pacific coast the soil is very productive and the climate delightful.

7. **The most important mountains** in the eastern part of North America are those of the Appalachian system, which comprises several ranges extending in the same general direction as the Atlantic coast.

8 CELEBRATED PEAKS IN NORTH AMERICA.

		Height in Feet.
Vol. Popocatapetl,	*Mexico*, highest	18,500
Mt. St. Elias,	*Alaska*	18,000
Mt. Whitney,*	*California*	15,086
Pike's Peak,	*Colorado*	14,500
Fremont's Peak,	*Wyoming*	13,570
Mt. Mitchell,*	*North Carolina*	6,707
Mt. Washington,	*New Hampshire*	6,428
Mt. Marcy,	*New York*	5,379
Mt. Mansfield,	*Vermont*	4,430

9. **From the base of the Rocky Mountains** to the Mississippi River, the descent is hardly perceptible, being a fall of only one foot in a mile.

10. **The great plains and lowlands** of North America extend from the Arctic Ocean to the Gulf of Mexico, and from the Rocky to the Appalachian Mountains.

11. **They comprise** four basins; those of the Mississippi, the St. Lawrence, Mackenzie River, and Hudson Bay.

Which of these basins are on the Atlantic slope? Which is on the Arctic slope?

12. **The Pacific Slope** of North America is that part which is west of the Rocky Mountains.

13. **It is drained** mainly by the Columbia, Colorado, and Yukon Rivers.

14. **From the Appalachian Mountains** to the Atlantic Ocean, the slope is gentle, being first hilly, then level, and near the coast, low and swampy. (See page 35.)

15. **The Atlantic Slope,** on account of its soil, streams, lakes, and inlets, is well adapted to agriculture, manufactures, and commerce.

LESSON X.

16. **The Climates** of North America are of every variety; tropical in the south, frigid in the north, and temperate in the middle.

17. **The climate of the Pacific Coast,** in the United States, British America, and Alaska, is much milder than that of the Atlantic coast, in corresponding latitudes. (See p. 11, par. 11.)

18. What countries of Europe are in the same latitude as Labrador—(see margins of the map)—as Newfoundland and Canada? Which extends furthest north, Lake Superior, New Brunswick, or France? At what part of the Old World would you arrive by sailing eastwardly from Halifax? From Cape May? From Savannah? From Florida Strait? What empires of Asia are directly west of San Francisco?

19. Winds partake of the temperature of the surface over which they blow. The land is warmer in summer and colder in winter than the ocean; that is, the temperature of the ocean is more uniform throughout the year than that of the land. And as the prevailing winds in the Temperate Zone are from the west, their temperature will be milder on the Pacific than on the Atlantic coast.

20. **The Rain** of the Pacific coast is supplied by vapor which rises from the Pacific Ocean.

21. On the Pacific coast of the United States the rain falls between that coast and the Sierra Nevada mountains, because the vapor is condensed before passing over the cold peaks of that high range, thus leaving a vast elevated region east of the Sierra Nevadas, from the Columbia to the Colorado river, destitute of rain. (See page 7, paragraph 41.)

* Mt. Whitney has nearly the same elevation as Mt. Blanc, the highest mountain in Europe. Mt. Mitchell is the highest peak of the Appalachian Mountains.

22. **The rain of the Atlantic Slope** is supplied from the Gulf of Mexico and the Atlantic Ocean.

23. **In the cold or northern regions** the vegetation is very scanty, while in the hot or tropical regions it is very dense, owing to the excessive heat and moisture.

24. **The principal animals** of the north are the white bear, reindeer, whale, walrus, and seal. In the south, alligators, turtles, and rattlesnakes are numerous; and among the wild animals of the temperate regions are bears, buffaloes, deer, and wolves.

25. **The Inhabitants** of the Arctic coasts are dwarfed in body and mind; in the tropical regions the inhabitants are enervated by the heat; but in the temperate climate man attains the highest degree of civilization.

26. **Greenland** is a cold, barren region, inhabited chiefly by Esquimaux, whose occupations are fishing and seal-hunting.

27. **Its European settlements** are on the western coast. Its eastern coast is constantly inclosed by ice.

28. **British America.**—Its northeastern half is covered with ice and snow nearly all the year. Its remaining part contains vast forests and prairies, where the buffalo, elk, deer, beaver, mink, and other animals, are hunted for their furs.

HISTORICAL GEOGRAPHY.

LESSON XI.

1. **America was discovered** by Christopher Columbus, in 1492 (October 12th). He landed on a small island, which he named San Salvador; and thinking it one of the islands of India, he called the natives Indians.

2. **Columbus** was an Italian; but he sailed under the orders of the King and Queen of Spain.

3. In the 9th century, Northmen from Norway colonized Iceland; and in the following century, Greenland was settled by Norwegian Icelanders.

4. **The mainland** of North America was discovered by John Cabot and his son, who sailed in the service of England, in 1497. The next year, his son conducted a voyage and explored the coast from Labrador to the Delaware Bay.

5. **America received its name** from Amerigo Vespucci (*ah-may-re'go ves-put-che*), who visited South America, in 1499.

6. Accounts of his voyages to the New World were published in Europe, and he acquired the reputation of being the first discoverer.

7. **Among the navigators** of the 16th century were Ponce de Leon, who explored Florida in search of a fountain which, according to a report, could restore youth to the aged; Cortes, who entered Mexico, which he found rich in gold and silver, and whose inhabitants (Aztecs) practiced many of the arts of civilization; and Balboa, who, from the Isthmus of Panama, discovered the Pacific Ocean.

8. **In the 17th century,** Henry Hudson, with the hope of reaching Asia, entered the bay which now bears his name.

9. **Among the celebrated voyagers** in the Arctic regions were Cook, Fox, Ross, Barrow, Parry, Sir John Franklin, and Kane; the most recent are Hall and Hayes.

10. **Nearly the whole of the New World** came into the possession of Spain and England, by right of discovery; but changes have been made by revolution, conquest, and treaties.

11. **Greenland and Iceland** belong to Denmark.

12. **Mexico and Central America** formerly belonged to Spain, but they are now independent.

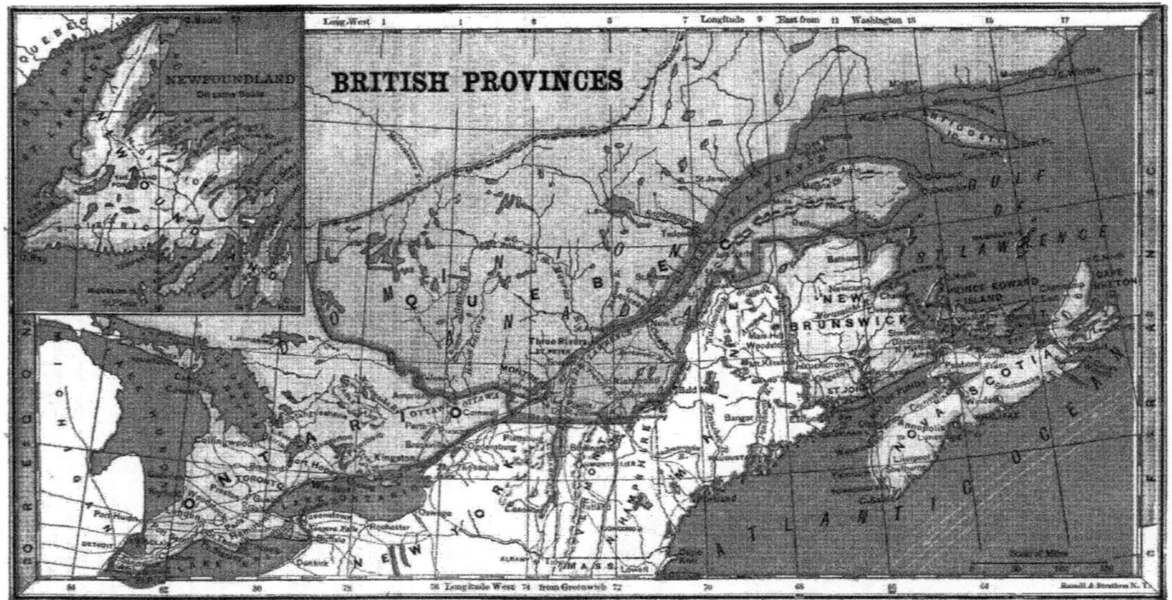

GENERAL QUESTIONS.

Name the British Provinces. What is the principal river? Through what province does the St. Lawrence flow? Between what Province and State does it flow? Mention three large tributaries on the north. On the south.

What bay between New Brunswick and Nova Scotia? What large river flows into it?

Of what lake is the Sorel' or Richelieu River the outlet?

Mention the principal cities in Quebec—Ontario—New Brunswick—Nova Scotia. Bound New Brunswick;—Ontario.

(See margins of the map.) What country in Europe is directly east of the British Provinces? What State on the Pacific coast is directly west?

Is Paris north or south of the city of Quebec? (See margin of page 23.) What three cities in Europe are in nearly the same latitude as Toronto?

REVIEW QUESTIONS.

Where are they?

Rivers:—St. Lawrence, Ottawa, Saguenay, St. Francis, Sorel', St. John's.

Bays:—Fundy, Chaleur, Georgian.

Cities and Towns:—Montreal, Quebec', Toronto, Hamilton, St. John's, Kingston, OTTAWA, Annapolis.

On what waters would you sail from Detroit to Buffalo? From Detroit to Toronto? From Toronto to Montreal? From Montreal to the Gulf of St. Lawrence?

COMPARATIVE SIZES.

The oblong here shown in dotted lines represents the size and shape of Kansas. It is used throughout this book as a *common measure*.

Observe that from north to south New Brunswick is the same as Kansas—200 miles; and the distance between the western part of New Brunswick and the eastern part of Cape Breton Island is equal to the length of Kansas—400 miles. (See p. 28.)

LESSON XIII.

THE DOMINION OF CANADA.

1. **The Dominion of Canada** comprises the provinces of Ontario, Quebec, New Brunswick, Nova Scotia, British Columbia, Manitoba, and Prince Edward's Island. The other **British Province** is Newfoundland. (See p. 16.)

2. **Its central part** is in the same latitude as the northern boundaries of Maine and Michigan.

3. **The winters** in the Province of Quebec are very severe, and last more than six months of the year.

4. **The temperature** of the atmosphere is the same as that of Norway, Sweden, and Iceland.

5. **The Climate** of Canada is hotter in summer and colder in winter than on the western coasts of the United States and Europe, in corresponding latitudes. (See page 11, paragraphs 10, 11, and 12.)

6. **Nearly all its rivers** flow into the St. Lawrence, and its northern boundary is the watershed which separates the Basin of Hudson Bay from that of the St. Lawrence River.

7. **The immense Forests** of British America furnish valuable timber, which,with furs andwheat, forms the chief export.

8. **The people of the eastern provinces** are largely engaged in lumber trade, ship-building, and the fisheries (cod, salmon, herrings, and mackerel).

NOTE.—Words in parentheses need not be committed to memory.

9. **By means of the Great Lakes** and the St. Lawrence River, commerce is extensively carried on between the interior of the continent and the Atlantic coasts of America and Europe.

Navigation around Niagara Falls and the rapids in the St. Lawrence is conducted by means of canals.

10. **Montreal** is the largest city in British America.

Quebec.—The St. Lawrence River.—View from Point Levi, looking north.

11. **Quebec**, situated on the top and at the foot of a promontory, is the most strongly fortified city in America, and the outlet for the products of Canada.

12. **Halifax** has one of the finest harbors in the world, and is a station for steamers sailing between Boston and England.

13.

BRITISH PROVINCES.	CAPITALS.	CHIEF CITIES.
ONTARIO	*Toronto*	Toronto.
QUEBEC	*Quebec*	Montreal.
NOVA SCOTIA	*Halifax*	Halifax.
NEW BRUNSWICK	*Fredrickton*	St. John.
NEWFOUNDLAND	*St. John's*	St. John's.*
PRINCE EDWARD'S ISLAND	*Charlottetown*	Charlottetown.
BRITISH COLUMBIA	*New Westminster*	Victoria.
MANITOBA.	*Winnipeg*	Winnipeg.

A Scene on the St. Francis River.—The logs are thrown into the streams, and floated down to the St. Lawrence River.

HISTORICAL GEOGRAPHY.

1. **Canada was colonized** by the French, under Cartier (in 1541).

2. **The Indians** were friendly until the French carried off one of their kings.

3. **Upper and Lower Canada,** now the Provinces of Ontario and Quebec, remained in possession of the French for more than 200 years, or until the capture of Quebec by English troops under General Wolfe (in 1759).

4. **Newfoundland and Nova Scotia** were ceded to England (in 1713). Nova Scotia means New Scotland.

5. **The People** of the province of Quebec are chiefly of French descent; of the other provinces, British.

6. **Each Province** has a legislature, which is elected by the people; the common parliament and governor-general are in Ottawa, the capital of the Dominion; but all are subject to Great Britain.

* The most eastern sea-port in North America.

Bound the United States on the north—east—south—west. On which side is the longest coast line? What capes on the Atlantic coast? The Pacific coast?

Through what Territories do the Rocky Mountains extend?

What mountains in California? Oregon and Washington? What chains in the eastern part of the United States?

What large river rises in Minnesota and flows south? What tributaries of the Mississippi rise in the Rocky Mountains? Which is the largest tributary from the east? What rivers flow into the Atlantic Ocean? Into the Gulf of Mexico? Into the Ohio River? What rivers are west of the Rocky Mountains?

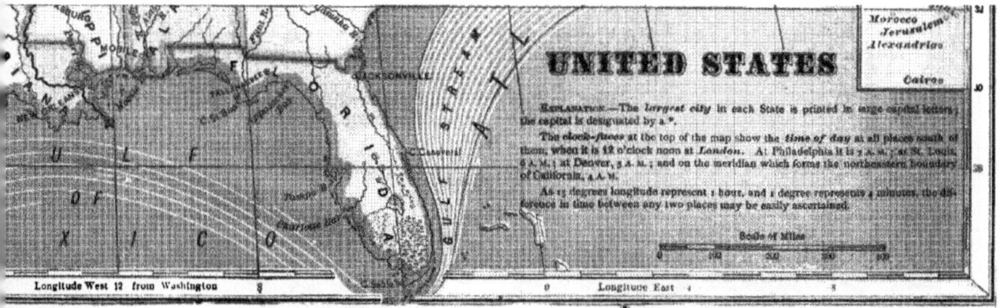
UNITED STATES
EXPLANATION.—The *largest city* in each State is printed in large capital letters; the capital is designated by a *.
The *clock-faces* at the top of the map show the *time of day* at all places south of them, when it is 12 o'clock noon at *London*.
As 15 degrees longitude represent 1 hour, and 1 degree represents 4 minutes, the difference in time between any two places may be easily ascertained.
Scale of Miles
Morocco
Jerusalem
Alexandria
Cairo
JACKSONVILLE
MOBILE
GULF STREAM
Longitude West 12 from Washington
Longitude East

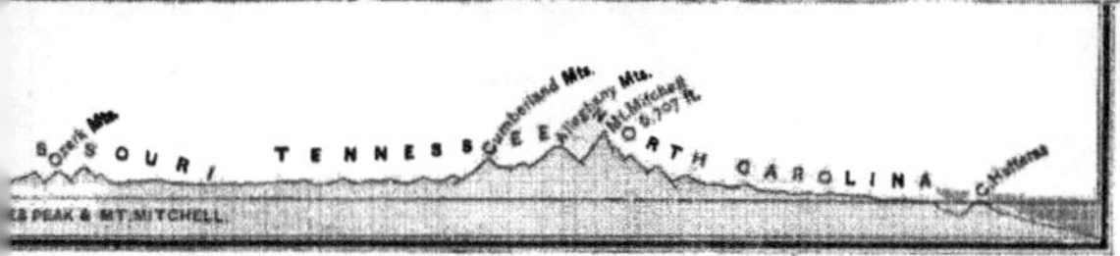
Ozark Mts.
Cumberland Mts.
Allegany Mts.
Mt. Mitchell 6,707 ft.
TENNESSEE
NORTH CAROLINA
C. Hatteras

...tion all the States north of the Delaware Bay which border on the At... ...cean. Mention those which are south of that bay. What States border ...ulf of Mexico? What are they called? *Ans. The Gulf States.* What ...order on the Pacific Ocean?

...at eight States border on the Great Lakes? What five States are on the west side of the Mississippi? What five are wholly on its east bank? What two States lie on both sides of the Mississippi?

COMPARATIVE LATITUDE AND TIME.

What large cities in Europe are further north than Quebec, Maine, and Michigan? What cities in the United States are in, or nearly in, the same latitude as Lisbon? Cairo? Pekin? Naples? Nice? Rome? What capital cities are on or near the parallel of 40° north latitude?

When it is noon at London, what o'clock is it at New Orleans? Savannah? Washington? New York? Boston? What is the difference in time between Chicago and San Francisco? Chicago and New York? Portland in Maine and Portland in Oregon?

Agriculture on the prairies.

PHYSICAL AND DESCRIPTIVE GEOGRAPHY.

1. **THE UNITED STATES** are situated in the North Temperate Zone, in the central part of North America, and between the same parallels of latitude as Southern Europe, the Mediterranean Sea, Northern Africa, Central Asia, and Japan.

Parallel 49°.

United States.	Southern Europe. Mediterranean Sea. Northern Africa.	Central Asia.	Japan.

Parallel 25°

2. **In the middle** of the North Temperate Zone of North America is Minnesota, which is equally distant from the Atlantic and Pacific Oceans.

3. **The most northern part** of the United States is on the parallel of 49 degrees, between Lake Superior and the Pacific Ocean.

4. **The most southern parts** are in Florida and Texas.

5. **The surface** is divided by the Rocky and the Alleghany Mountains into three great sections: the Pacific Slope, west of the Rocky Mountains; the Atlantic Slope, east of the Alleghanies; and the Mississippi Basin, between them. Besides these, are the Gulf Slope, the basin of the great lakes and the St. Lawrence River, and the basin of the Red River of the North.

6. **The western half** of the United States comprises high mountains and plains; the eastern half is mostly level or undulating, except the Appalachian System of Mountains, extending from Georgia to the Gulf of St. Lawrence.

7. **The two high ranges** are the Rocky and the Sierra Nevada, between which are extensive table-lands remarkable for their aridity and barrenness.

Elevation of highest peaks, about 15,000 feet; of table-lands, 4,000 to 6,000 feet.

8. **This country** possesses nearly every variety of climate, owing to its great extent, its position on the globe, and differences in elevation.

9. **Climate varies** according to latitude, elevation, and the influences of the ocean, winds, and mountain ranges.

10. **In the north and northeast,** the winters are long and severe; the summers, short and hot.

11. **In the south,** the summers are long and hot; and the winters, mild.

12. **Ascending the high peaks** of the Rocky Mountains and the Sierra Nevadas, the traveler finds the cold to increase, and reaches the limit of perpetual snow. (See page 11, paragraph 9.)

13. **Compared with Western Europe,** the climate of the greater part of the United States is warmer in summer, colder in winter, and dryer.

14. **Rain** is well distributed over the States. The largest quantity falls on the Pacific, the Gulf, and the Atlantic States; and the least, on the great table-lands which extend from the Sierra Nevadas eastwardly into Western Kansas and Northern Texas.

15. **Snow** lies from three to five months in the most northern States, but it seldom falls south of Virginia, except among the mountains.

16. **The States remarkable** for their agricultural products are those in the eastern half of the Union; grain, fruits, and vegetables in the north, and cotton, tobacco, rice, and sugar in the south.

17. **The prairie land** of Ohio, Indiana, Illinois, Michigan, Wisconsin, Iowa, Missouri, Arkansas, Kansas, and Nebraska, is remarkable for its fertility. These include most of the Central and Lake States.

18. **Celebrated for wheat, corn, and wool,** are the Lake and Central States, and California.
Cotton, the South Atlantic and the Gulf States.
Tobacco, the Central States.
Cane Sugar, Louisiana.
Rice, South Carolina.

19. **The States and Territories remarkable for precious metals** are between the Rocky Mountains and the Pacific Ocean.

Gold, California, Colorado, Montana, and Idaho.
Silver, Nevada.
Quicksilver, California.

20. **Coal and the useful metals** abound in many of the States and Territories which lie between the Rocky Mountains and the Atlantic Ocean.

21. **Celebrated for coal and iron** is the region of the Alleghany Mountains.

22. **Coal and iron** are extensively mined in Pennsylvania;
Lead, in Michigan, Wisconsin, Illinois, and Iowa;
Iron and lead, in Missouri.

Cultivation of the sugar-cane in Louisiana.

Interior of a cotton-mill.—In the distance are seen steamships engaged in commerce.

17. **The leading manufacturing States** are in the northeastern part of the Union.

18. **The principal manufactures** are cotton and woolen goods, flour, machinery, iron and steel ware, boots, shoes, and leather.

19. **The cotton and woolen manufactures** of all the New England States, except Vermont, are very extensive.

20. **Flour and lumber** are largely produced in nearly every State in the Union.

21. **The commerce** of the United States is very important.

22. **The principal commercial States** are New York and Massachusetts.

23. **The principal ports** of foreign commerce are New York and Boston in the north, New Orleans in the south, and San Francisco in the west; of inland commerce, are St. Louis, Chicago, Cincinnati, Buffalo, Cleveland, Detroit, and Milwaukee.

HISTORICAL GEOGRAPHY.

1. **The first settlements** in the United States were formed about a century and a quarter after the discovery of America.

2. **A Spanish colony,** St. Augustine, was founded in Florida, in 1565.

3. **English colonies** were founded at Jamestown, Virginia, in 1607, and at Plymouth, Massachusetts, in 1620.

4. **A Dutch colony** was founded on Manhattan Island, now the city of New York, in 1613.

5. **A Swedish colony** was founded in Delaware in 1638.

6. **These settlements** came gradually under the control of the English, who organized the Thirteen Colonies of

NEW HAMPSHIRE, MASSACHUSETTS, RHODE ISLAND,
CONNECTICUT, NEW YORK, NEW JERSEY,
PENNSYLVANIA, DELAWARE, MARYLAND,
VIRGINIA, GEORGIA, NORTH CAROLINA
AND SOUTH CAROLINA.

7. **When the English had governed the colonies** about a century and a quarter, the colonists declared themselves free and independent, on the Fourth of July, 1776.

8. **The war of the Revolution,** which began in 1775, arose chiefly from unjust taxation of the American colonies by England.

9. **The government** of this country, under its Constitution and its first President George Washington, began in 1789, since which Florida and all the land west of the Mississippi River have been acquired by the United States.

10. **The Northwest Territory** in 1787 comprised Ohio, Indiana, Illinois, Michigan, and Wisconsin.

11. **The governments** in this Republic comprise the general, State, and Territorial governments.

12. **Each comprises** three branches—the legislative, executive, and judicial.

13. **The legislative power** of the general government is vested in Congress; the executive power, in the President; and the judicial power, in the Supreme and certain other courts.

14. **Congress** is composed of the Senate and House of Representatives.

15. **The Senate** is composed of two Senators from each State, elected for six years.

16. **The House of Representatives** is composed of members elected every two years from the several States, according to the population.

17. **The Vice-President** presides over the Senate, and in the event of the President's death, resignation, or removal, he becomes President.

18. **The Constitution provides** that representation and direct taxation shall be in proportion to the number of the inhabitants of the several States.

19. **Every bill, to become a law,** must be passed by both houses of Congress and signed by the President; if he disapprove, the bill must be reconsidered and passed by two-thirds of each house.

20. **The United States shall guarantee** to every State a republican form of government, and protect each from invasion.

21. **Amendments to the Constitution** may be made on application of the legislatures of two-thirds of the States, and when ratified by the legislatures of three-fourths of the States.

22. **Congress shall make no law** respecting an establishment of religion, or prohibiting the free exercise thereof.

The United States Senate.

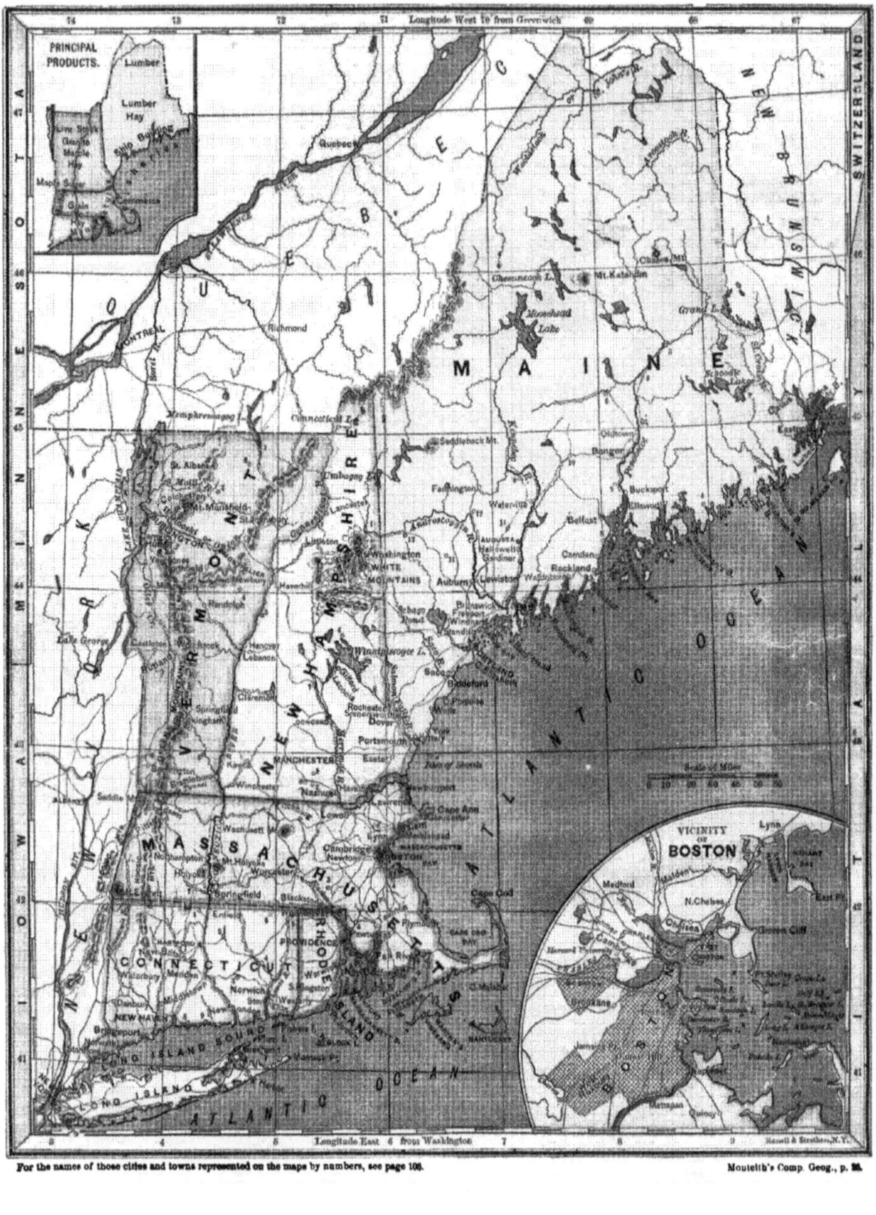

For the names of those cities and towns represented on the maps by numbers, see page 108.

LESSON XIX.

THE NEW ENGLAND STATES

1. **The New England States** are situated near the central part of the North Temperate Zone—in the northeastern part of the United States—and in the same latitude as Italy, the northern part of the Chinese Empire, Oregon and Michigan.

2. **Their area** is about equal to that of Missouri.

3. **Three of the six border** on Canada, and all but one have sea-coast.

GENERAL QUESTIONS.

In New England, what range of mountains? What group of mountains? Which is the largest river? Between what two and through what two States does it flow? What rivers flow into the Atlantic Ocean? Into Long Island Sound?

What large lake west of Vermont? What lake in New Hampshire? What is the outlet of Lake Champlain? What rivers flow into Lake Champlain? What countries in Europe are directly east of the New England States? (See margin of large map.) What Western States are directly west?

Bound Maine. What is its capital? Its metropolis? (See "Cities and Towns," with note in "Review.") What are its sea-ports? Bound New Hampshire What is its capital? Its metropolis? Its leading sea-port? Bound Massachusetts. What city is its capital and metropolis? What are its sea-ports? Bound Rhode Island. What are its capitals? Its sea-ports? What is its metropolis? Bound Connecticut. What are its capitals? Its sea-ports? What is its metropolis? Bound Vermont. What is its capital? Its metropolis? Has Vermont any sea-ports?

LESSON XX.

DIRECTIONS FOR DRAWING THE STATES.

All the States are drawn on the *same Unit of Measure.*
Each measure represents 200 miles.
Mark all names which appear in large type.

TO DRAW MAINE.

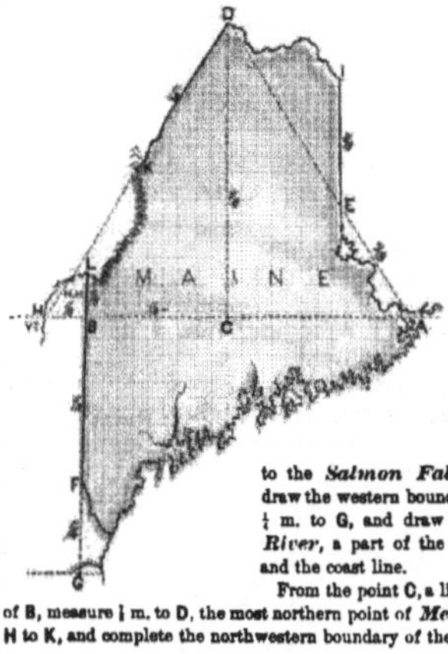

Make a scale like the one shown below, on a slip of stiff paper; and by it, draw maps of all the States, as directed.

Draw no lines except boundaries.

Commence at A, draw *Passamaquoddy Bay*, and locate *Eastport.* Measure west, on the 45° of latitude, *one measure* to B; thence ⅛ m. to L, the northeast corner of *N. H.*, and ⅛ m. to H, the northeast corner of *Vt.* From B measure ½ m. to the *Salmon Falls River* at F, and draw the western boundary. From F measure ¼ m. to G, and draw the *Salmon Falls River*, a part of the *Merrimac River*, and the coast line.

From the point C, a little less than ½ m. east of B, measure ¾ m. to D, the most northern point of *Me.*; thence ½ m. towards H to K, and complete the northwestern boundary of the State.

From A measure ¾ m. toward D, to E, and draw *Grand Lake* and *St. Croix River.* From E measure north to I ¾ m., and draw the eastern boundary line and the *River St. John.*

Next, draw from the large map, in the following order, with their names: the mountains,—the rivers,—the bays,—the capes,—the cities and towns—(mark those only which appear on the large map in ***black*** letters); then mark the railroads. In drawing a map of your own State, mark all the cities and towns.

Write in each State, its principal products, or the leading occupations of its inhabitants.

REVIEW QUESTIONS.

In what part of the State is each of the following?

Mountains:—Mt. Katah'din, Saddleback Mt., Chase's Mt.
Rivers:—Penobscot, St. Croix (*kroy*), Androscoggin, Kennebec, Aroos'took, Woolastook, Saco (*saw'ko*).
Lakes:—Moosehead, Chesun'cook, Schoodic, Umba'gog, Grand.
Bays:—Penobscot, Casco, Fundy, Frenchman's.
Cities:—Portland, Bangor, **Lewiston,** Biddeford, AUGUSTA, Bath,

NOTE.—The metropolis or largest city in the State is at the head of the list, and the capital is in capital letters.

A fine forest in Maine.

1. **Maine is situated** exactly midway between the Equator and the North Pole.

2. **In size** it is equal to Scotland, or to the five other States of New England.

3. **It is noted** for its broken coast line, rugged surface, vast evergreen forests, numerous lakes and streams, long and severe winters.

4. **It excels** every other State in shipbuilding.

5. **Its principal cities and towns** are in the southern part of the State. **Portland** is its largest and chief commercial city. **Bangor** carries on a flourishing trade in lumber.

6. **Its principal slope** is southward to the Atlantic coast.

SCALE FOR DRAWING ALL THE STATES.

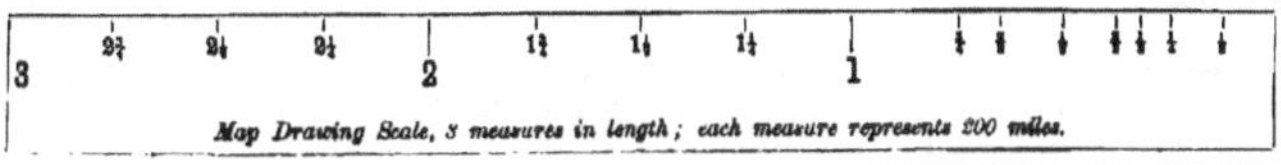

Larger maps may be drawn easily by increasing the length of the measure.

The White Mountains of New Hampshire.—View from Mt. Washington.

LESSON XXI.

NEW HAMPSHIRE AND VERMONT.

MAP-DRAWING.

The measurements for the eastern boundary of New Hampshire are the same as those for the western boundary of Maine. Make a scale like the one given at the foot of page 27.

Begin at **A**, measure ½ m. north to **B**, ½ m. south to **J**, ½ m. from **J** to **L**, and ½ m. east from **L** to **K**. Complete the eastern boundary of the State, drawing ***Salmon Falls River***, the ***Atlantic Coast***, and ***Cape Ann***. Measure ½ m. from **L** to **H**, and ½ m. from **H** to **C**. Draw ***Merrimac River***, and complete the southern boundary of the States.

From **A**, measure ½ m. west to **C**, ½ m. from **C** to **E**, ½ m. from **E** south to **F**, and ½ m. from **F** to **G**. Draw ***Lake Champlain, Connecticut River***, the ***Green*** and ***White Mountains***.

Complete the drawing by adding, from the large map, the rivers,—the bays,—the capes,—the cities and towns,—the railroads.

NOTE.—Here the drawings will be examined by the teacher, or the pupils will draw the map on the blackboard, each doing a part.

REVIEW QUESTIONS.

What is the situation of each of the following?

Mountains:—Green, White, Mt. Washington, Mt. Mansfield.

Rivers:—Connecticut, Merrimac, Otter Creek, Onion or Winooski, Salmon Falls.

Lakes:—Champlain, Winnipiseogee (*-saw'ke*), Memphremagog.

CITIES AND TOWNS.

NOTE.—The Metropolis of each State stands first; the Capital is in capital letters.

New Hampshire.	Nashua,	**Vermont.**
Manchester,	Portsmouth,	Burlington,
CONCORD,	Dover.	MONTPELIER.

COMPARATIVE SIZES, LATITUDES, ETC.

IN AREA.

Vermont and New Hampshire, same as Greece, 19,000 square miles.

IN POPULATION.

Maine, New Hampshire, Vermont, Massachusetts and Connecticut, combined, same as London, 3½ millions.

IN LATITUDE.

Concord, same as Marseilles (France) . . . . . 43°

IN HEIGHT.

Mt. Washington (6,400 ft.), less than one-half of Mt. Blanc (15,000 ft.).

LESSON XXII.

MASSACHUSETTS, CONNECTICUT, AND RHODE ISLAND

MAP-DRAWING.

Begin at **H**, and measure ½ m. to **G** and ½ m. to **K**, and draw the northern boundary of the State, the ***Merrimac River*** and ***Cape Ann***.

Draw the western boundary, ½ m. from **G** to **P**. From **P**, measure ½ m. a little west of south to **R**, and draw the western boundary of ***Connecticut*** and the western end of ***Long Island***. From **P**, measure ½ m. to **N**, ½ m. from **N** to **M**, ½ m. from **M** to **L**, ½ m. from **L** to **O**, and ½ m. from **O** south to **S**. Draw the northern boundary of ***Connecticut*** and ***Rhode Island***; ***Plymouth Bay, Cape Cod, Cape Cod Bay***, and the eastern coast of ***Massachusetts***; ***Martha's Vineyard*** and ***Nantucket***.

Draw the boundary line between ***Connecticut*** and ***Rhode Island***, ½ m., a little west of **N**.

Draw the northern shore of ***Long Island Sound, Narragansett Bay***, and ***Buzzard's Bay***.

Complete these States by adding the mountains, rivers, principal towns, and railroads.

Draw ***Long Island***, observing that its eastern extremity extends to a point directly south of the eastern boundary of ***Connecticut***.

What is the length, in miles, of the western boundary of Rhode Island? Of the western boundary of Massachusetts?

If the teacher wish, the pupils will now draw the six Eastern States together.

REVIEW QUESTIONS.

Where are they?

Mountains:—Hoosic, Saddle, Everett, Holyoke, Wachu'sett.

Rivers:—Blackstone, Housatonic (*hoo-sa-ton'ik*), Thames (*temz*).

Bays:—Cape Cod, Buzzard's, Narragansett.

CITIES AND TOWNS.

Massachusetts.	Lawrence,	**Rhode Island.**	HARTFORD,
BOSTON,	Lynn,	PROVIDENCE,	Bridgeport,
Worcester,	New Bedford,	NEWPORT.	Norwich,
Lowell,	Fall River,	**Connecticut.**	Norwalk,
Cambridge,	Springfield.	New Haven,	Waterbury.

COMPARATIVE SIZES AND LATITUDES.

IN AREA AND POPULATION.

Massachusetts, Connecticut, and Rhode Island, together, are the same as Switzerland, 14,000 sq. m.,—2,300,000 inhabitants.

IN LATITUDE.

Plymouth, same as Rome (Italy) . . . . . 42°

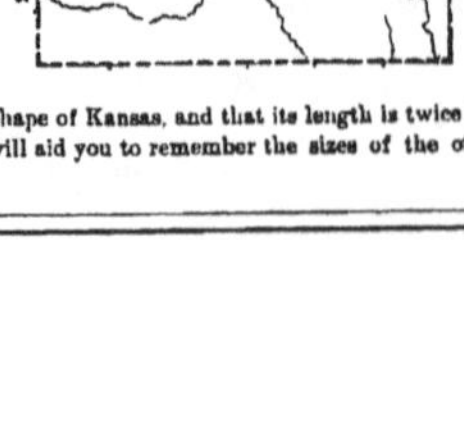

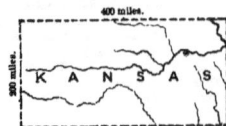

Observe the shape of Kansas, and that its length is twice its width—200 by 400 miles. It will aid you to remember the sizes of the other States. (See next page.)

PHYSICAL AND DESCRIPTIVE GEOGRAPHY.

LESSON XXIII.

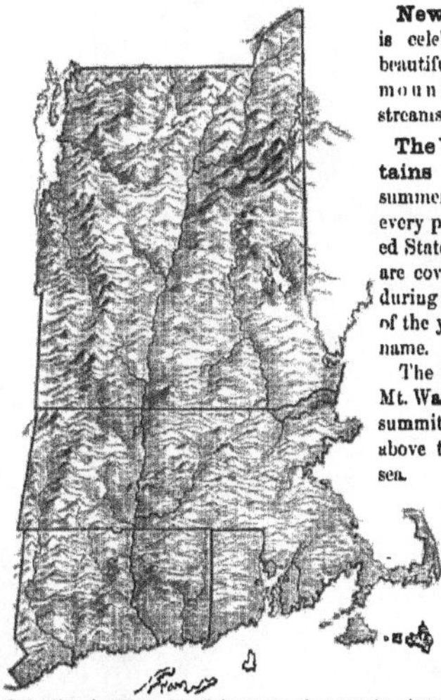

Relief Map showing the face of the country—the mountains, rivers, river valleys, and low coast land.—Point out the mountains—lakes—capes. Follow the course of each of the large rivers from the source to the mouth.

New Hampshire is celebrated for the beautiful scenery of its mountains, lakes, streams, and waterfalls.

The White Mountains are visited by summer tourists from every part of the United States. Their peaks are covered with snow during the greater part of the year, hence their name.

The highest peak is Mt. Washington, whose summit is over 6100 ft. above the level of the sea.

In what part of the State are the White Mountains? What rivers have their sources in them? What large river of Maine flows past them? What large river rises in the northern part of New Hampshire? What can you say of the Connecticut? *Ans. It is the largest river in New England.*

New Hampshire is one of the leading States in the manufacture of cotton and woolen goods.

The three large cities of New Hampshire are Manchester, Concord, and Nashua. Concord is noted for the manufacture of carriages and the quarrying of granite.

Vermont is also remarkable for its mountains, fine scenery, and evergreen forests.

Its valleys are fertile, and its hills afford excellent pasture for sheep, horses, and cattle. It produces large quantities of wool, butter, cheese, and maple sugar.

Its largest city is Burlington, which is noted for its beautiful situation and its facilities for trade. Where is it situated?

Massachusetts excels every other State in the Union, in the manufacture of cotton and woolen goods, leather, boots and shoes.

Its eastern part is celebrated for large manufacturing towns; and its western, for the delightful summer resorts among its hills and valleys.

Mention eight of the largest cities in Massachusetts, and give the situation of each. (See List of Cities on page 4.) What range of mountains in the western part of the State? The highest peak in the State is in the northwest—what is its name? What other peaks in the State?

Boston is one of the leading commercial cities in the Union, and has an excellent system of Public Schools. Harvard University is in Cambridge.

On the Merrimac River, both in Massachusetts and New Hampshire, are large manufacturing cities: mention them.

Connecticut is also largely engaged in manufactures, especially those of woolen and cotton goods and hardware. It excels every other State in the Union in the manufacture of silk goods, India-rubber goods, fire-arms, and sewing machines.

Its surface slopes toward the south, and its rivers flow into Long Island Sound.

Its river valleys are celebrated for their fertility and beauty.

Mention its principal river. What river in the western part of the State? Eastern? Mention the largest cities in Connecticut. (See page 4.)

New Haven, called the "City of Elms," is one of the handsomest cities in the Union. Next in rank are Hartford, the capital, and Bridgeport, an important manufacturing city. Yale College is in New Haven.

In what part of the State is New Haven? Hartford? Bridgeport?

Rhode Island, the smallest State in the Union, manufactures more cotton goods than any other State, except Massachusetts.

Its capitals are Providence and Newport, which are also the largest cities in the State.

HISTORICAL GEOGRAPHY.

LESSON XXIV.

1. **The first permanent settlement** in New England was made at Plymouth, Massachusetts, by the Pilgrim Fathers (English), December 22, 1620.

2. **The colonists** suffered much from cold, hunger, and the hostility of the Indians.

3. **The English colonies** in America belonged to Great Britain for more than a century, or until the Revolutionary war, which began in 1775; the cause of the war was taxation, without representation.

4. **The first skirmish** occurred at Lexington, Massachusetts, between the colonists and the British troops (April, 1775).

5. **The first important battle** in the Revolution was that of Bunker Hill, which occurred on the 17th of June, 1775.

6. **Among the first thirteen States** were Massachusetts, New Hampshire, Connecticut, and Rhode Island; the first added to the original thirteen States was Vermont.

7. **The soldiers of Vermont** who fought in the Revolution were called the Green Mountain Boys.

8. **The people of New England** are mostly of English descent.

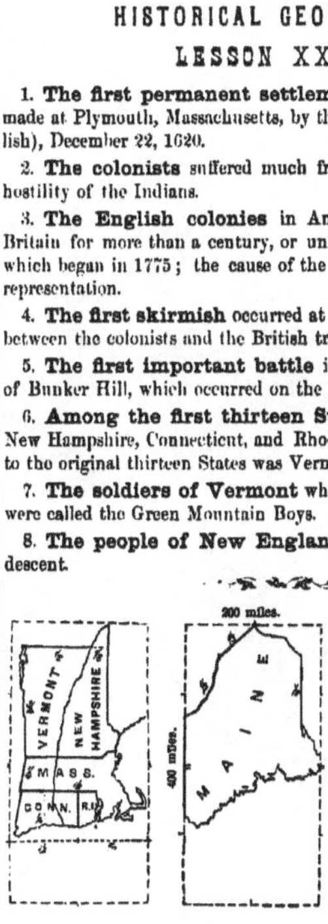

To impress more firmly upon the mind the **Comparative Sizes** of the States, all are drawn upon **Kansas as a Common Measure.**

OBSERVE that five of the New England States may be drawn inside of Kansas, that the width of *Maine* is the same as that of *Kansas*—200 miles; and that *Massachusetts*, from east to west, is almost 200 miles.

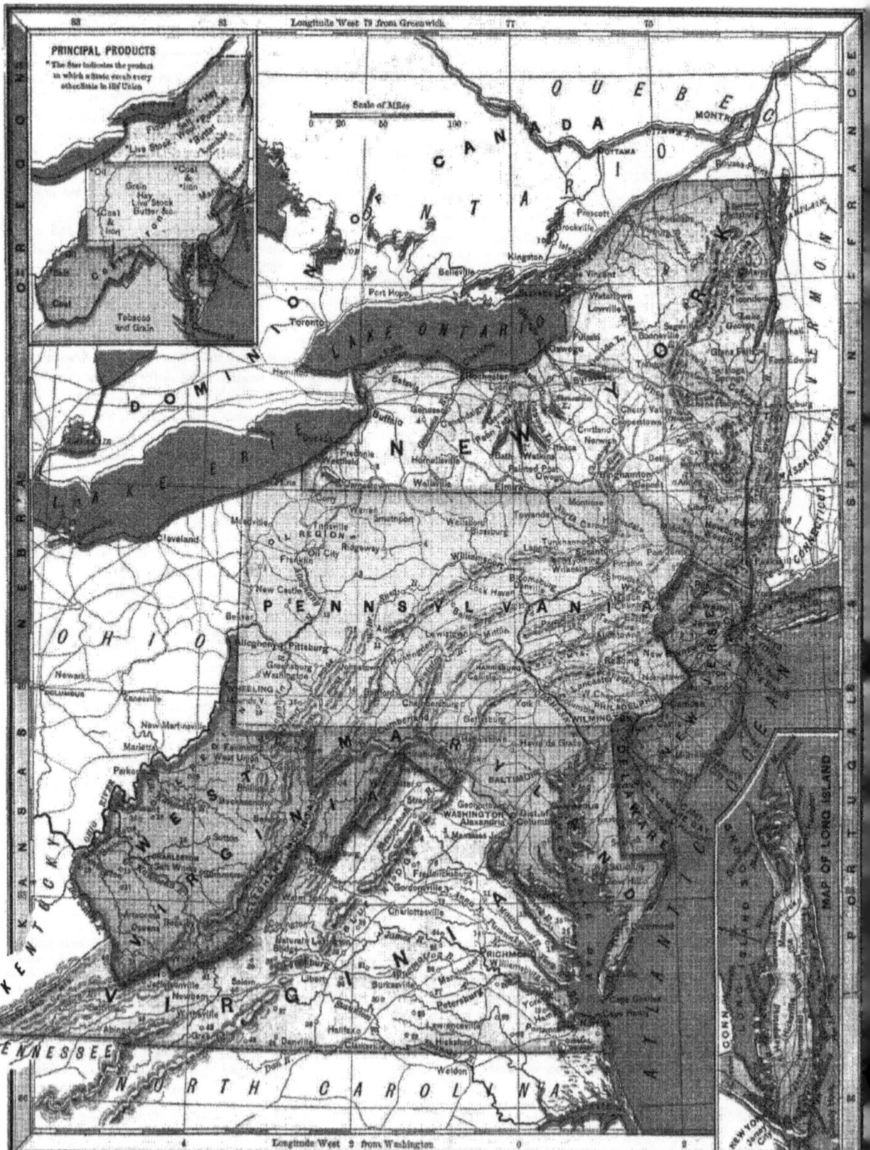

For the names of places represented by numbers, see p. 103.

THE MIDDLE ATLANTIC STATES.

LESSON XXV.

1. The **Middle States** comprise New York, Virginia, and the intervening States.

2. **They lie** between the parallels $36\frac{1}{2}°$ and 45° north latitude, and are in the North Temperate Zone.

3. **The northern boundary** of New York is exactly midway between the Equator and the North Pole.

4. **The principal watershed** is formed by the Alleghany and Adirondack Mountains. East of it the rivers flow toward the Ocean; west of it, into the Great Lakes, the Ohio and St. Lawrence Rivers.

GENERAL QUESTIONS.

Name the States on this map. Which of them border on the Atlantic Ocean? On Lake Erie? Which one on Lake Ontario? What Lake northeast of New York? What two large bays extend into these States? What large river empties into the head of Delaware Bay? Of Chesapeake Bay?

What two capes at the mouth of Delaware Bay? Of Chesapeake Bay?

What mountain ranges in New York? Pennsylvania? Virginia and Maryland? In what direction do they extend? In what direction does the coast extend?

TO DRAW NEW YORK.

(For scale, see foot of page 27.)

Begin at **E**, measure $\frac{1}{2}$ m. south to **F**, $\frac{1}{4}$ m. from **F** to **G**, $\frac{1}{2}$ m. from **G** to **P**, and $\frac{1}{2}$ m. a little west of south to **R**. Draw ***Lake Champlain***, the eastern boundary of the State, ***Long Island***, ***Long Island Sound***, ***Staten Island***, ***Sandy Hook***, and the southern extremity of ***Hudson's River***. Locate ***New York***, ***Brooklyn***, and ***Jersey City***.

From **P** measure $\frac{1}{2}$ m. west to **B**. From **B** measure $\frac{1}{4}$ m. toward **R** and draw a part of ***Delaware River*** and the northern boundary of ***New Jersey*** $\frac{1}{4}$ m.

From **B** measure a little more than $1\frac{1}{4}$ m. west to **K**, $\frac{1}{4}$ m. north from **K** to **L**, and $\frac{1}{4}$ m. east from **L** to **M**, and draw the northern boundary of ***Pennsylvania*** and the eastern extremity of ***Lake Erie***. Locate ***Buffalo*** and ***Niagara Falls***.

From **L** measure $\frac{1}{4}$ m. north to **N**, and 1 m. east from **N**. Draw ***Lake Ontario***, observing that it is $\frac{1}{4}$ m. wide.

From **E** measure $\frac{1}{3}$ m. west to **D**, and $\frac{1}{8}$ m. from **D**, toward **K**, to **S**, and draw ***St. Lawrence River***. Complete the State by marking the mountains, rivers, cities, railroads, and the Erie Canal.

LESSON XXVI.

REVIEW QUESTIONS.

What rivers flow into the St. Lawrence River? Into Lake Ontario? What is the outlet of the lakes in the center of the State? What lake in the eastern part of the State empties into Lake Champlain? Into what does Lake Champlain empty?

In what mountains does the Hudson rise?

Mention the principal cities and towns (those in capital letters) in the northern part of the State—in the eastern—in the western—in the central part.

Bound New York.

Where are they? (See Note, page 15.)

Mountains:—Adirondack, Catskill, Highlands, Mt. Marcy.

Rivers:—Hudson, Mohawk, Genesee, Chenango, Black, Racket, Oswego, Saranac.

Lakes:—Ontario, Erie, Champlain, George, Oneida, Cayuga, Seneca, Otsego.

Cities:—New York, Brooklyn, Buffalo, ALBANY, Rochester, Troy, Syracuse, Utica.

PHYSICAL AND DESCRIPTIVE GEOGRAPHY.

1. **NEW YORK** is in the middle of the North Temperate Zone, and in the same latitude as Oregon and Southern France.

2. **In the east and southeast** are mountains with rich valleys and pasture lands; in the centre and west, the land is mostly level and very productive.

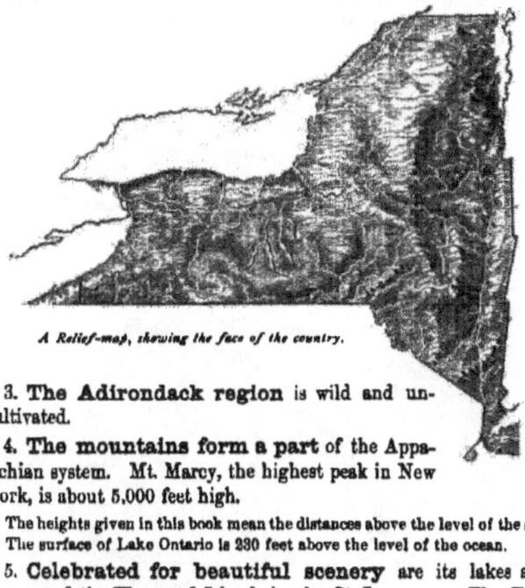

A Relief-map, showing the face of the country.

3. **The Adirondack region** is wild and uncultivated.

4. **The mountains form a part** of the Appalachian system. Mt. Marcy, the highest peak in New York, is about 5,000 feet high.

The heights given in this book mean the distances above the level of the sea. The surface of Lake Ontario is 230 feet above the level of the ocean.

5. **Celebrated for beautiful scenery** are its lakes and rivers, and the Thousand Islands in the St. Lawrence. The Falls of Niagara are noted for their grandeur.

6. **New York**, called the Empire State, excels in population, wealth, and commerce.

LESSON XXVII.

7. **The largest city** in the Western Continent is New York, which is admirably situated for commerce.

8. **Among its advantages** are its fine harbor, and its communications with other States, by rivers, railroads, and canals.

9. **Its population** is about one million, and is equal to that of Constantinople, or one-half that of Paris, or one-third that of London.

Philadelphia and St. Petersburg have each a population of about 700,000.

Cities which have about 400,000 inhabitants are Brooklyn, St. Louis, Chicago, Naples (the largest in Italy), and Rio Janeiro (the largest in South America).

10. **Brooklyn** is the third city in the Union, in population.

11. **Finely situated** for manufactures and inland trade are Buffalo, at the eastern extremity of Lake Erie; Rochester, at the falls in the Genesee River; and Oswego, at the mouth of the Oswego River.

12. **Troy** is at the head of steamboat navigation on the Hudson.

13. **Syracuse** has valuable salt works.

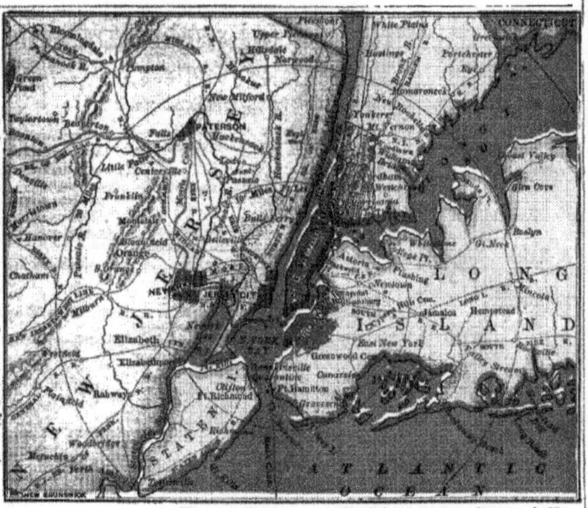

COMPARATIVE SIZES, ETC.

New York State is the same, in area, as Portugal and Belgium combined,—47,000 square miles.

New York State has the same population as Sweden—over 4 millions.

Hudson River (300 miles) is in length ½ that of the Rhine.

New York City is in the same latitude as Naples,—almost 41°.

Draw a circle around Albany, at a distance from it of 100 miles, and observe that it passes near New Haven and Rome, and exactly over the northern point of New Jersey.

HISTORICAL GEOGRAPHY.

1. **When New York was first entered** by white men, it was occupied by warlike tribes of Indians, called the Five Nations.

2. **Champlain,** a French navigator, first entered the State from Canada, by way of the lake which bears his name (1609).

3. **Henry Hudson,** an English navigator, discovered New York Bay, and entered the river which received his name (1609). He was in search of a passage across the continent. (See page 19, paragraph 8.)

4. **The land which Hudson discovered** was claimed by Holland, and named New Netherlands, because he was in the employ of the Dutch East India Company.

5. **The first settlements** formed by the Dutch were at Fort Orange, now Albany, and New Amsterdam, now New York City.

6. **The right of the Dutch** to New Netherlands was disputed by the English, because of the prior discovery by Cabot. (See page 19, paragraph 4.)

7. **New Netherlands** was captured by the English, who changed the name to New York (1664), after the Duke of York.

LESSON XXVIII.

PENNSYLVANIA AND NEW JERSEY.

MAP-DRAWING.

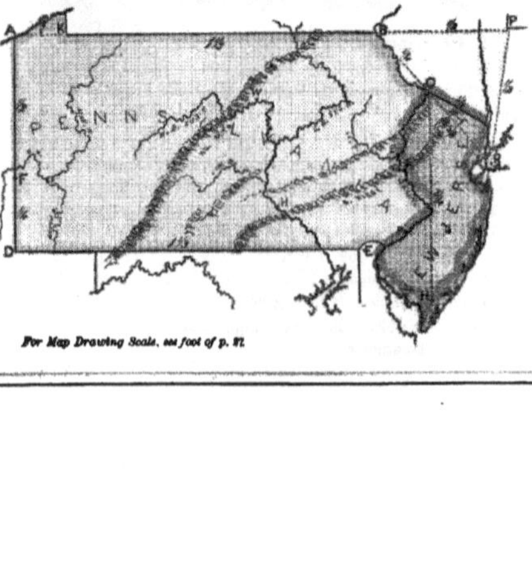

For Map Drawing Scale, see foot of p. 27.

Begin at B, measure 1½ m. west to A, and a little less than ½ m. from A to K. Draw the northern boundary line, and a part of the shore of ***Lake Erie.*** Locate ***Erie City.*** From A, measure south ¾ m. to D, and draw a part of the ***Ohio River***, near F, ½ m. south of A.

Measure 1½ m. east from D to E, and draw the southern boundary line of ***Pennsylvania***, including the northern line of ***Delaware.***

Complete the outline of the State by drawing the ***Delaware River.*** Locate the northern corner of ***New Jersey,*** ¼ m. from B towards R, at O. Locate ***Brooklyn,*** ¾ m. from B, at R. Draw ***Staten Island.*** Locate ***Jersey City*** and ***New York.*** Draw a part of ***Hudson River*** and the northern boundary of ***New Jersey,*** ½ m. Measure ¾ m. south of O to H, and draw the eastern shore-line of the State and ***Delaware Bay.***

Draw the mountains and rivers. Locate the principal capes, cities, towns, and railroads.

REVIEW QUESTIONS.

Name the largest rivers in these two States. In what States are their sources? Into what does each empty? What two meet and form the Ohio River? What two cities at their junction? On what two rivers is Philadelphia?

In what part of Pennsylvania is the coal region? The oil region?

What is the southern cape of New Jersey? Mention the principal cities in New Jersey. Mention those in Pennsylvania which are between the Delaware and Susquehanna. Mention those in the western part of Pennsylvania.

Bound Pennsylvania—New Jersey.

In what part of the State are they?

Rivers:—Susquehanna, Monongahela, Delaware, Alleghany, Schuylkill, Juniata.

Cities:—Philadelphia, Allegheny, Newark, Scranton, Pittsburg, HARRISBURG, Jersey City, TRENTON.

LESSON XXIX.

PHYSICAL AND DESCRIPTIVE GEOGRAPHY.

Relief Map of Pennsylvania, showing the face of the country.

1. **PENNSYLVANIA** is remarkable for its mountains and the abundance of its coal, iron, and rock-oil.
2. **Its mountain ranges** and valleys extend in a northeasterly direction.
3. **Its manufactures** are numerous, especially those of iron, cotton, and wool.
4. **The soil** of its valleys is very productive.
5. **Philadelphia**, the largest city in the State, is the most important manufacturing city in the Union.

Relief Map of New Jersey.

6. **Pittsburg** is celebrated for its iron works and coal trade.
7. **NEW JERSEY** is level in the south and center, and hilly and mountainous in the north.
8. **The soil**, except in the south and east, is well adapted to grazing and agriculture.
9. **Newark**, the metropolis of New Jersey, is a flourishing manufacturing city; it is on the Passaic River.

COMPARATIVE SIZES, ETC.

New Jersey, same area as the Kingdom of Wurtemberg (Germany), 8,000 square miles.

Pennsylvania, same population as Holland (3½ millions).

Pittsburg, same latitude as Madrid (Spain), $40\frac{1}{2}$°.

Draw a circle around Philadelphia, having a radius of 100 miles. It will pass over the northern point of New Jersey and near the southern boundary of Delaware. Over or near what cities will it pass?

HISTORICAL GEOGRAPHY.

1. **PENNSYLVANIA** was settled by the Swedes and Finns (natives of Sweden and Finland) in 1643, and afterward by William Penn and other Quakers from England. (*Sylva*, a wood.)
2. **Unlike other colonies**, it was long free from trouble with the Indians, owing to the just treatment which they received.
3. **Memorable events** which occurred in Pennsylvania are the defeat of Braddock, near Pittsburg, during the French and Indian war; the battles of Brandywine and Germantown, the winter encampment at Valley Forge, and the meeting of the first Congress, during the Revolution.
4. **The Declaration of Independence** was adopted in Philadelphia, July 4th, 1776.
5. **NEW JERSEY was first settled** by the Dutch (1620), and, with New Amsterdam and Delaware, passed under the control of the English (1664).
6. **Important victories** were won in New Jersey by the Americans during the Revolution, at Trenton, Princeton, and Monmouth.

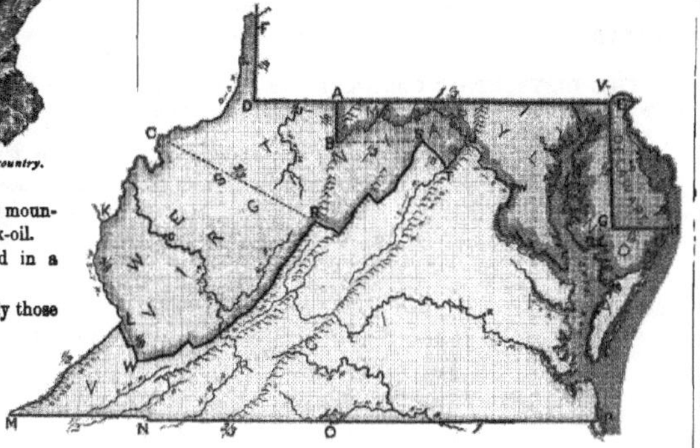

LESSON XXX.

VIRGINIA, WEST VIRGINIA, MARYLAND, AND DELAWARE.

MAP-DRAWING.

Begin at D and measure ¼ m. north to F, and draw the ***Pan Handle.*** Locate ***Wheeling.*** Measure from D to E, and draw the northern boundaries of ***West Virginia***, ***Maryland***, and ***Delaware.*** Next, draw the western and southern boundaries of ***Delaware*** ⅓ m. from V to G and ¼ m. from G to H. Draw ***Delaware Bay*** and locate ***Dover***, ***Capes May*** and ***Henlopen.*** Mark A ⅓ m. east of D, and draw the western boundary of Maryland ⅛ m. from A to B.

Next mark the point P 1½ m. south of E, and draw ***Chesapeake Bay*** and the ***Potomac River.*** Locate ***Washington*** and ***Baltimore***, ***Capes Charles*** and ***Henry.***

Draw the southern boundary line of ***Virginia*** 1⅓ m. from P to N, and ⅓ m. from N to M. From M measure ⅓ m. northeast to L, and draw the ***Cumberland Mountains.*** Measure north ⅔ m. from L to K, and draw the ***Big Sandy River.*** Draw the ***Ohio River*** from F to K.

Complete the eastern boundary of ***West Virginia***, observing that the southern point of the State is at W ½ m. southeast of L; that the breadth of the State is ⅗ m. from C to R; and that the point S is ⅓ m. east of B.

Complete the map by marking the ***Mountains***, ***Rivers***, etc.

LESSON XXXI.

REVIEW QUESTIONS.

Name the rivers of Virginia—of West Virginia. In what directions do they flow? Which are boundary rivers? Name the chief cities in Virginia—in West Virginia—in Maryland—in Delaware. Bound Delaware—Maryland—Virginia—West Virginia.

In what direction from Washington is Richmond? Baltimore? Wheeling? Oswego? What large cities are almost in a straight line between Washington and New Haven?

What capes at the mouth of Chesapeake Bay? Delaware Bay?

Where are they situated?

Rivers;—Potomac, Monongahela, Great Kanawha, James, Shenandoah, Rappahannock, Little Kanawha.

Cities:—Baltimore, WASHINGTON, RICHMOND, Wilmington, Petersburg, Norfolk, WHEELING, Charleston.

PHYSICAL AND DESCRIPTIVE GEOGRAPHY.

1. **DELAWARE** is level; its principal agricultural productions are wheat and peaches.

2. **Its chief city** is Wilmington, which is celebrated for its manufactures; its car-works are the largest in the United States.

3. **MARYLAND** is level in the east, and mountainous in the west.

4. **THE DISTRICT OF COLUMBIA**, on the Maryland shore of the Potomac, contains Washington, the capital of the United States.

5. **VIRGINIA** is mountainous in the west; its surface slopes toward the Chesapeake Bay; its valleys, especially that of the Shenandoah, are noted for their fertility.

6. **WEST VIRGINIA** is mountainous in the east; its surface slopes toward the Ohio River.

7. **The principal products** of Maryland, Virginia, and West Virginia are coal, iron, tobacco, and grain; and their chief cities—Baltimore, Richmond, and Wheeling—are celebrated for their manufactures and commerce. Baltimore is one of the largest cities in the Union.

COMPARATIVE SIZES, ETC.

Virginia is a little larger than Portugal.

Richmond has about the same population as Quebec, 52,000.

If you draw a circle of 100 miles radius around Washington, over or near what three large cities will it pass?

HISTORICAL GEOGRAPHY.

1. **The first English settlement** in the United States was Jamestown, in Virginia (1607).

2. **The settlers** suffered much from famine, disease, and the hostilities of the Indians.

3. **In the French and Indian war**, Washington, then a young man, distinguished himself; by his skill and bravery, he saved Braddock's army (English) from ruin (1755).

4. **In the Revolution**, Virginia contributed largely to the success of the American Union.

5. **The Presidents** of the United States who were natives of Virginia were Washington, Jefferson, Madison, Monroe, Harrison, Tyler, and Taylor.

6. **West Virginia** formed a part of Virginia until 1863.

7. **Maryland** was settled by emigrants from England (1634); Delaware, by Swedes and Finns (1638).

States compared with each other, by placing them on KANSAS, AS A COMMON MEASURE OR FRAME, which is here represented by the oblongs enclosed in dotted lines. (*See foot of page 29.*)

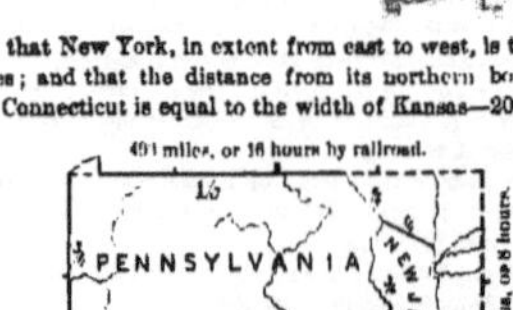

Observe **that New York, in extent from east to west, is the same as Kansas—400 miles; and that the distance from its northern boundary to Pennsylvania and Connecticut is equal to the width of Kansas—200 miles.**

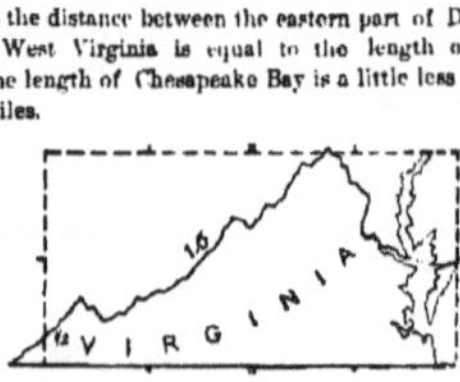

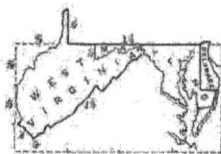

Observe that the distance between the eastern part of Deleware and the western part of West Virginia is equal to the length of Kansas—400 miles; and that the length of Chesapeake Bay is a little less than the width of Kansas—200 miles.

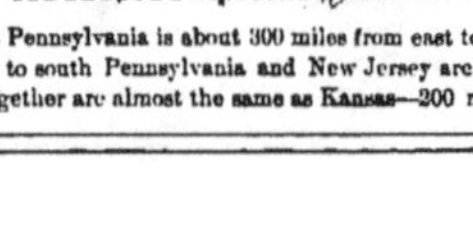

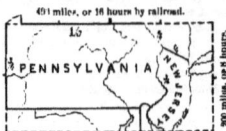

Observe that Pennsylvania is about 300 miles from east to west; that in extent from north to south Pennsylvania and New Jersey are equal to each other, and both together are almost the same as Kansas—200 miles.

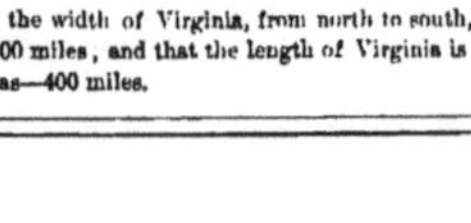

Observe that the width of Virginia, from north to south, is the same as that of Kansas—200 miles, and that the length of Virginia is but little more than that of Kansas—400 miles.

RELIEF MAP

Of a part of the United States, showing the face of the country between the Mississippi River and the Atlantic Ocean—the Appalachian Mountain System, the Atlantic Slope, and part of the Mississippi Basin.

The pupils will point to the mountain ranges which compose the Appalachian system, and trace each large river from its source to its mouth, observing the valley which is drained by it.

Observe the slope from the mountains to the Atlantic and the Gulf of Mexico, and the low, swampy land near the coasts; also, that midway between the mountains and the coast there is a sudden descent of the surface. At this line are waterfalls in the rivers, also the head of navigation; consequently, cities and towns were built here on account of manufacturing and commercial facilities. On this line are Richmond, Petersburg, Weldon, Raleigh, Columbia, and Augusta.

Monteith's Comp. Geog., p. 38.

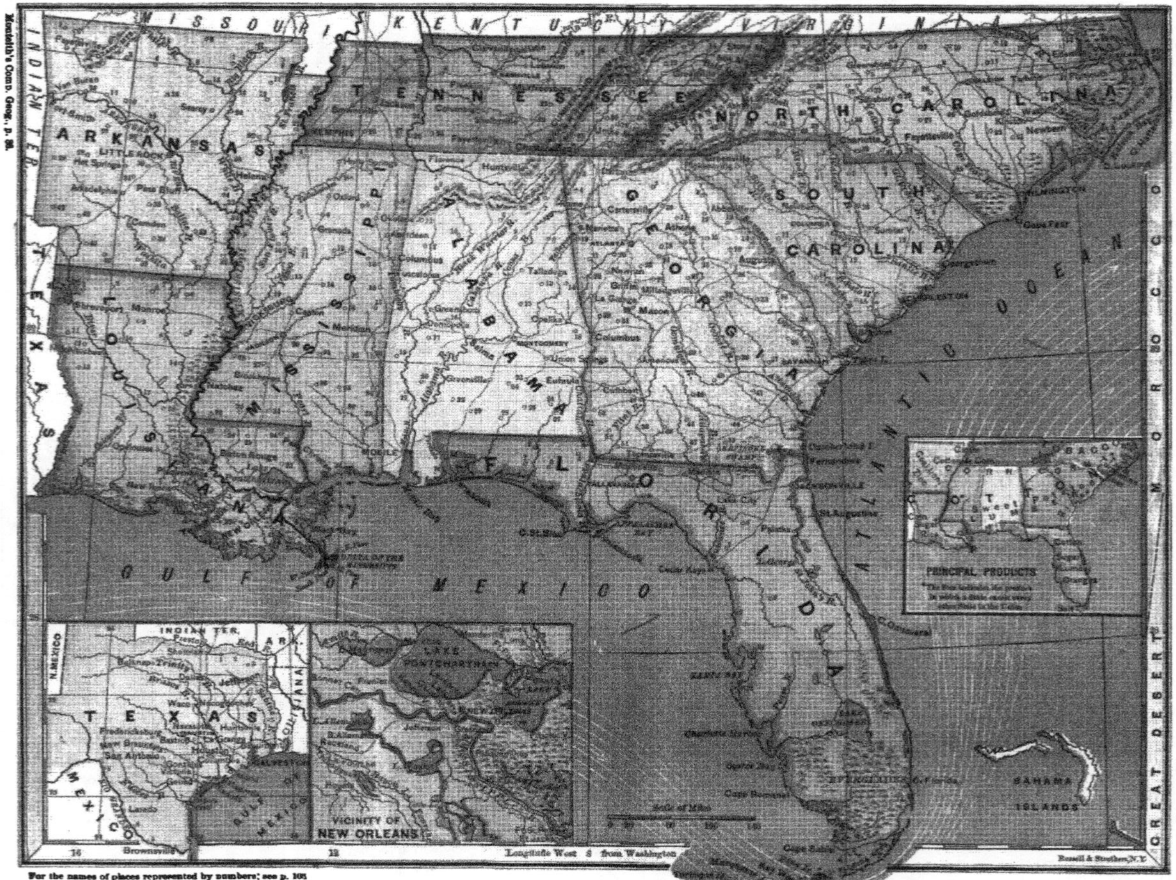

For the names of places represented by numbers; see p. 105

THE SOUTHERN AND SOUTHWESTERN STATES.

1. **They comprise** the Gulf States, with N. Carolina, S. Carolina, Georgia, Tennessee, and Arkansas.

2. **The Gulf States** are Florida, Alabama, Mississippi, Louisiana, and Texas; they are in the southern part of the North Temperate Zone, and constitute the warmest section of the United States.

3. **The northern boundaries** of North Carolina, Tennessee, and Arkansas, almost coincide with the parallel of 36½° north latitude.

4. **On account of their warm, moist climate** and fertile soil, the Southern States are rich in agricultural products, the most important of which are cotton, corn, tobacco, sugar-cane, rice, and sweet potatoes.

GENERAL QUESTIONS.

What part of Africa is directly east of these States? (See margin.) What part of Asia is directly west? Which of these States border on the Atlantic Ocean? On the Gulf of Mexico? On the Mississippi River?

What ranges of mountains have their southern extremities in these States? Which of these States have no mountains?

Mention the boundary rivers. What rivers flow into the west side of the Mississippi River? Into the east side?

What rivers in North Carolina? In South Carolina? In Georgia? In Florida? In Alabama? In Mississippi? In Louisiana? In Arkansas?

What bays and sounds on the coast of North Carolina? What three capes? Are these capes on the main-land or islands?

What bays and capes on the west coast of Florida? What capes on the east coast?

Draw a map OF NORTH CAROLINA, SOUTH CAROLINA, and GEORGIA.

REVIEW QUESTIONS.

What cities in these three States, on or near the coast? What sounds east of North Carolina? Bound North Carolina—South Carolina—Georgia. Mention their capitals.

Where are they? (See Note, page 15.)

Rivers:—Roanoke, Neuse, Altamaha' (haw'), Santee, Great Pedee, Congaree, Chattahoochee, Oco'nee, Savannah.

Cities:—Charleston, Savannah, Augusta, Columbus, RALEIGH, Wilmington, ATLANTA, COLUMBIA, Macon.

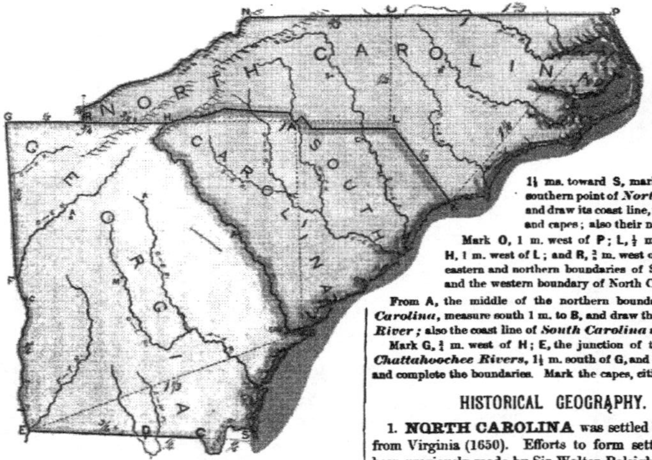

PHYSICAL AND DESCRIPTIVE GEOGRAPHY.

1. **Mountains** are in the western part of North Carolina and the northwestern parts of South Carolina and Georgia.

2. **The surface of these States** slopes south-easterly to the Atlantic Ocean, with an abrupt descent midway between the mountains and the coast. *(See relief map on page 35.)*

3. **From their center** to the coast, the land is level and sandy, producing cotton, rice, and sweet potatoes.

4. **The pine forests**, especially of North Carolina, are very extensive, and furnish tar, pitch, turpentine, resin, and lumber.

5. **The hills and valleys** of the upper portions of these States are well adapted to the production of grass and grain.

6. **The largest city** in South Carolina is Charleston; in Georgia, Savannah; and in North Carolina, Wilmington.

7. **Georgia** has the same area as England and Wales.

Draw the northern boundary of *North Carolina*, 1¾ measures in length. Mark the mouth of *St. Mary's River*, at S, 2 ms. south of N. From P, 1½ ms. toward S, mark K, the most southern point of *North Carolina*, and draw its coast line, with its sounds and capes; also their names.

Mark O, 1 m. west of P; L, ½ m. south of O; H, 1 m. west of L; and R, ¾ m. west of H. Draw the eastern and northern boundaries of South Carolina and the western boundary of North Carolina.

From A, the middle of the northern boundary of *South Carolina*, measure south 1 m. to B, and draw the *Savannah River*; also the coast line of *South Carolina* and *Georgia*.

Mark G, ¾ m. west of H; E, the junction of the *Flint* and *Chattahoochee Rivers*, 1½ m. south of G, and 1 m. west of S, and complete the boundaries. Mark the capes, cities, etc.

HISTORICAL GEOGRAPHY.

1. **NORTH CAROLINA** was settled by emigrants from Virginia (1650). Efforts to form settlements had been previously made by Sir Walter Raleigh.

2. **The first settlement in Georgia** was made by the English, under Oglethorpe, where Savannah now stands.

3. **The colonists** of North and South Carolina and Georgia suffered much from Indian depredations, and afterward, in the Revolution, from invasion by the British troops.

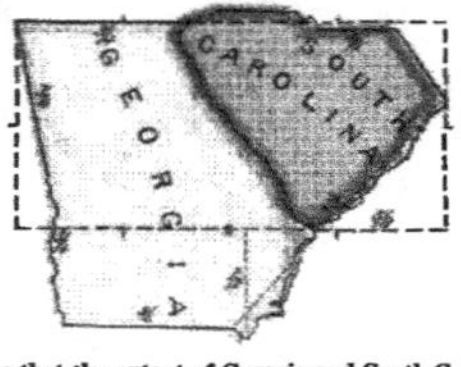

Observe that the extent of Georgia and South Carolina from east to west is the same as that of Kansas—400 miles; and that, from north to south, South Carolina is the same as Kansas—200 miles. (See pages 28 and 29)

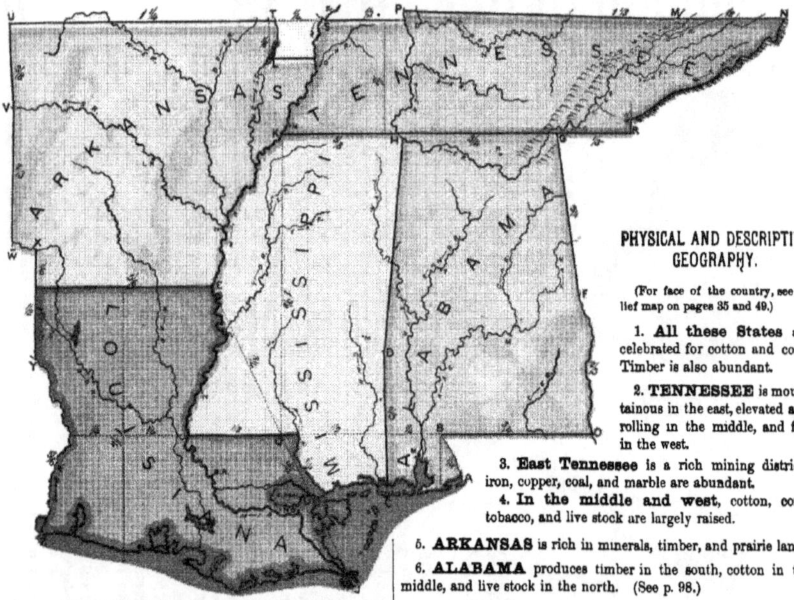

DIRECTION.—*The maps may be drawn* by the pupils at home, and examined by the teacher the next day; or, in the class-room, on their slates; or, in turn, on the blackboard.

MAP-DRAWING.

Begin at N, and draw the northern boundaries of Tennessee and Arkansas, according to the distances shown on the map.

½ m. south of P, mark H; also K, G, and R, and complete the boundaries of Tennessee, its mountains and rivers.

South of K, mark Q, then S and O, and complete the boundaries of Mississippi and Alabama.

Draw the western boundaries of Arkansas and Louisiana, beginning at U. The mouth of the Mississippi is in a line with E and Q.

Complete the boundaries, and add the mountains, rivers, chief cities, etc.

REVIEW QUESTIONS.

Which of these five States have no sea-coast? Which have no mountains? Which are partly bounded by the Mississippi? Mention the other boundary rivers. Mention each State, with its capital.

	Where do they rise? In what directions do they flow?	*Where do they empty?*	
Rivers:—Tennessee,	Cumberland,	Red,	Pearl,
Tombigby,	Arkansas,	Mobile,	White,
Yazoo,	Alabama,	Sabine,	Coosa.

Cities:— *Where are they?*

NEW ORLEANS,	NASHVILLE,	MONTGOMERY,	Baton Rouge,
Memphis,	Vicksburg,	Natchez,	Selma,
Mobile,	LITTLE ROCK,	Knoxville,	JACKSON.

PHYSICAL AND DESCRIPTIVE GEOGRAPHY.

(For face of the country, see relief map on pages 35 and 49.)

1. **All these States** are celebrated for cotton and corn. Timber is also abundant.

2. **TENNESSEE** is mountainous in the east, elevated and rolling in the middle, and flat in the west.

3. **East Tennessee** is a rich mining district; iron, copper, coal, and marble are abundant.

4. **In the middle and west**, cotton, corn, tobacco, and live stock are largely raised.

5. **ARKANSAS** is rich in minerals, timber, and prairie lands.

6. **ALABAMA** produces timber in the south, cotton in the middle, and live stock in the north. (See p. 98.)

7. **MISSISSIPPI** is noted especially for cotton, and Louisiana and Florida for sugar and tropical fruits—oranges, lemons, figs, etc.; these States have extensive forests and swamps.

8. **New Orleans**, the largest city in the Union south of Baltimore, is the principal cotton market in the south. During a freshet, the Mississippi is higher than the streets of the city, requiring dikes or levees to prevent inundation.

9. **Mobile**, the largest city in Alabama, is an important cotton market. Among the advantages of its situation are its means of communication by sea and rivers.

10. **The other large cities** in these States are Memphis, Nashville, and Natchez.

11. **New Orleans** is in the same latitude as Cairo (Egypt).

HISTORICAL GEOGRAPHY.

1. **ARKANSAS and LOUISIANA** were settled by the French, and formed part of the large tract called Louisiana, which extended to British America and the Pacific Ocean. The purchase of this tract from France gave to the United States full control of the Mississippi River.

2. **ALABAMA and MISSISSIPPI** were first visited by Spaniards, under De Soto (about 1540), in search of gold, without success. De Soto died, and was buried in the Mississippi River.

FLORIDA.

Draw the northern boundary to correspond with the southern boundaries of Georgia and Alabama.

What river forms the northeastern boundary of Florida? In what swamp does the St. Mary's River rise? Draw St. Mary's River and locate Okefenoke Swamp. What town in Florida opposite the mouth of the St. Mary's River? Is Fernandina on the mainland? Locate Fernandina. What river forms the northwestern boundary of Florida? A few miles east of its mouth is the largest city in the western part of Florida. Name and locate it.

What two rivers from Georgia meet on the northern boundary of Florida? What river is formed by them? Draw them. What town at the mouth of the Appalachicola River? Locate it, and draw the coast line between it and the western boundary of Florida.

Mark Cape Sable 2 ms. a little east of south from the mouth of St. Mary's River.

Mark N on Tampa Bay 1 m. south of C. Draw Tampa Bay and the coast line to Appalachicola and Cape Sable.

Mark Cape Canaveral ¾ m. northeast of Tampa Bay, and draw the coast line from the mouth of St. Mary's River to Cape Sable. Complete the map.

REVIEW QUESTIONS.

Where are they?

Capes:—Sable, Florida, Canaveral, St. Blas.

Rivers:—St. John's, Appalachicola, Suwanee.

Cities:—Jacksonville, Pensacola, Fernandina, TALLAHASSEE, St. Augustine, Appalachicola.

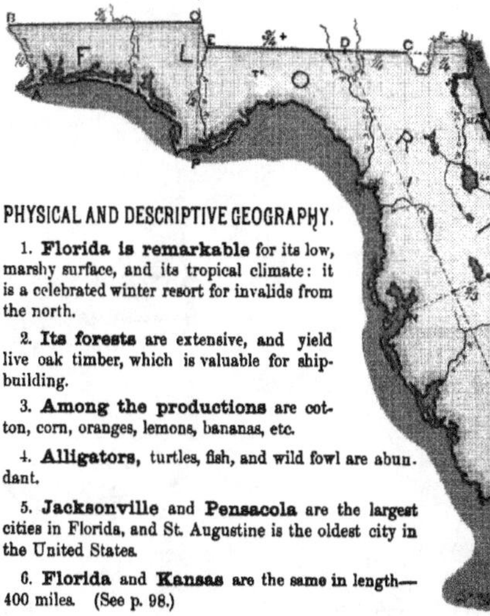

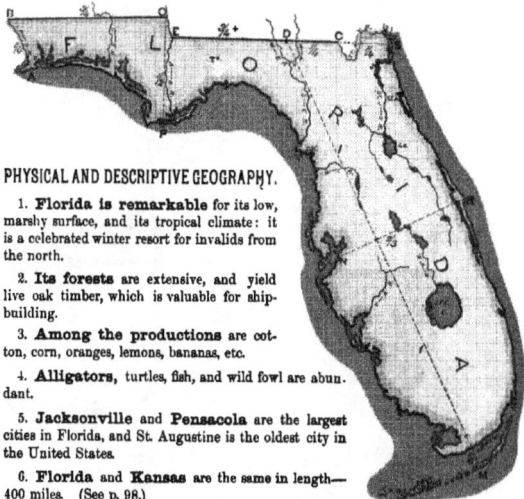

PHYSICAL AND DESCRIPTIVE GEOGRAPHY.

1. **Florida is remarkable** for its low, marshy surface, and its tropical climate: it is a celebrated winter resort for invalids from the north.

2. **Its forests** are extensive, and yield live oak timber, which is valuable for ship-building.

3. **Among the productions** are cotton, corn, oranges, lemons, bananas, etc.

4. **Alligators**, turtles, fish, and wild fowl are abundant.

5. **Jacksonville** and **Pensacola** are the largest cities in Florida, and St. Augustine is the oldest city in the United States.

6. **Florida** and **Kansas** are the same in length—400 miles. (See p. 98.)

7. **Florida** was settled by Spaniards (in 1565), and was purchased from Spain by the United States (1820).

THE LAKE AND CENTRAL STATES.

1. **The Lake and Central States** lie in the northern half of the Union, just east of its center. (See next page.)

2. **They border** on the Mississippi, the Ohio, and the Great Lakes.

3. **Mountains** are in the eastern part of Kentucky and southern part of Missouri, and high land in Minnesota.

4. **These States are remarkable** for their rich prairie land and their rapid growth in population and wealth.

5. **The people** are engaged mainly in agriculture, manufactures, and stock-raising.

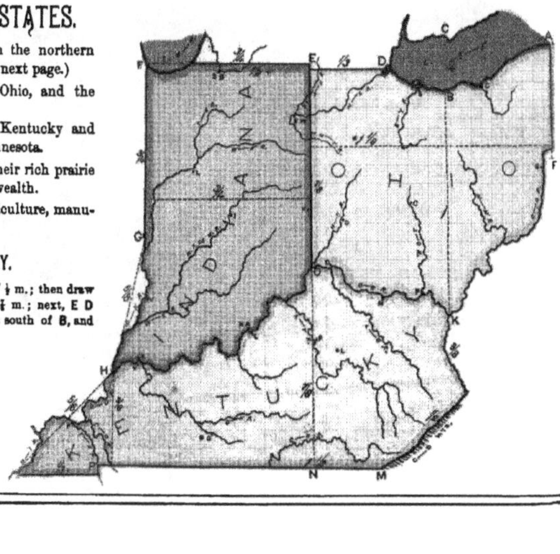

OHIO, INDIANA, AND KENTUCKY.

Draw the eastern boundary of ***Ohio*** from A to F ½ m.; then draw the western, 1½ m. west of the eastern, from E to O, ¾ m.; next, E D ½ m. and ***Lake Erie*** ½ wide. Find the point K 1 m. south of B, and draw the ***Ohio River***. Complete the State.

Join ***Indiana*** to ***Ohio*** by drawing its northern boundary with ***Lake Michigan*** ¾ m. from E to F; its western, ¾ m. from F to G; the ***Wabash River*** ¾ m. from G to H; and the ***Ohio River*** from O to H.

Draw ***Kentucky*** by measuring ¼ m. from O south to N. Mark the southern boundary 1½ m. from M to P, the ***Tennessee River***; and ½ m. from P to S, the ***Mississippi River***. Locate the principal rivers, mountains, etc., as in the other maps.

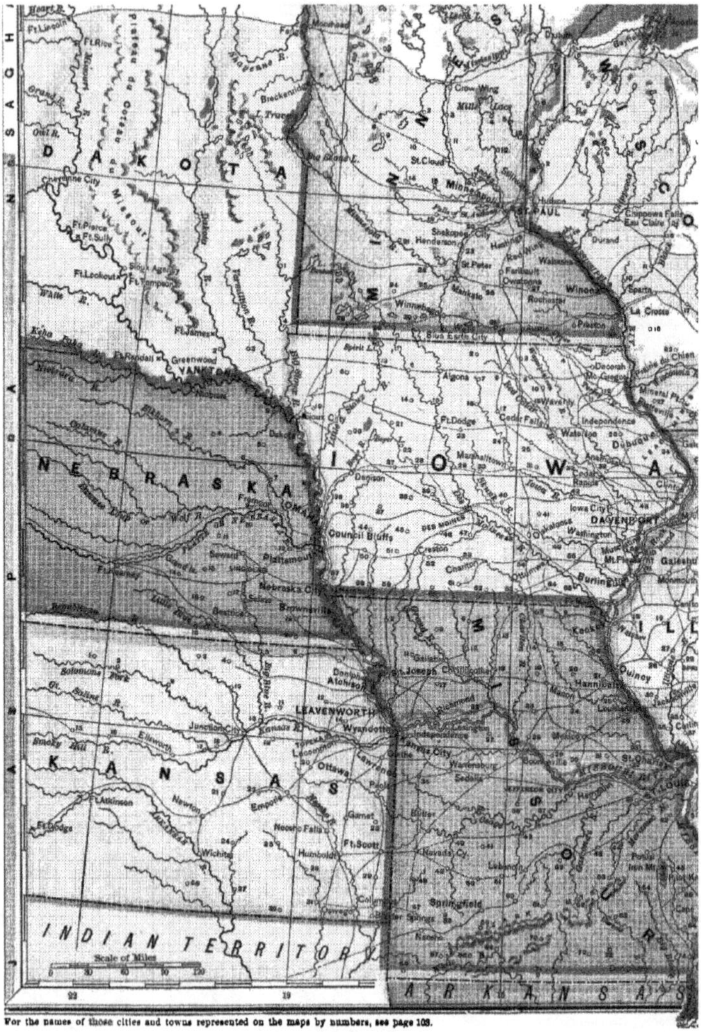

For the names of those cities and towns represented on the maps by numbers, see page 103.

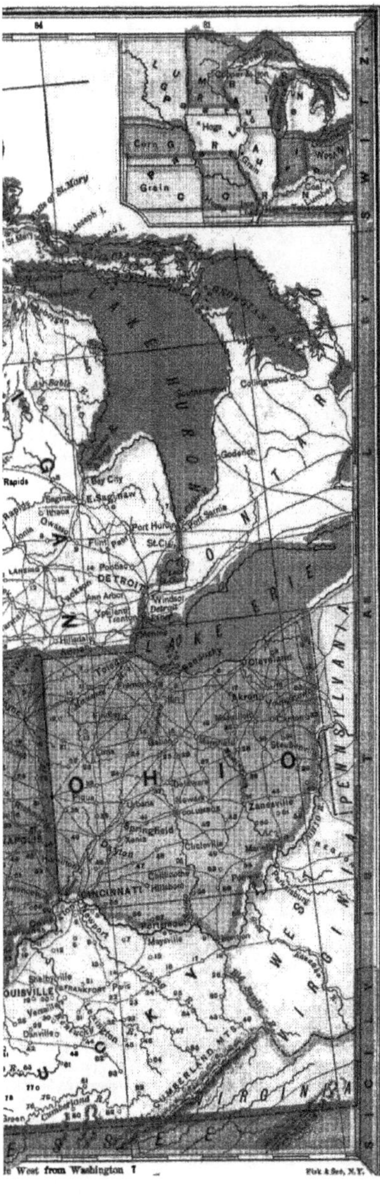

REVIEW QUESTIONS ON OHIO, INDIANA, AND KENTUCKY.

Mention the boundary rivers of Ohio, Indiana, and Kentucky. What rivers flow into the Ohio River? Into Lake Erie? Mention the largest cities on Lake Erie—on the Ohio River.

Where are they? (See Note, page 15.)

Rivers:—Ohio, Kentucky, Cumberland, Maumee, Wabash, Big Sandy, Scioto, Miami.

Cities:—Cincinnati, INDIANAPOLIS, Terre Haute, Newport, Louisville, COLUMBUS, Evansville, Lexington, Cleveland, Covington, Fort Wayne, New Albany, Dayton, Toledo, FRANKFORT, Sandusky.

PHYSICAL AND DESCRIPTIVE GEOGRAPHY.

1. **OHIO and INDIANA** are rich in their agricultural and manufactured products; their trade, which is very extensive, is by way of the lakes, the Ohio River, canals and railroads.

2. **They produce** corn, wheat, hay, and wool, in abundance.

3. **KENTUCKY** is high in the southeast, and low in the north and west. It excels every other State in the culture of tobacco.

4. **Its forests** produce excellent timber; and on its grassy tracts, horses, mules, and cattle are extensively raised.

5. **Coal** is abundant in Kentucky, Ohio, and Indiana.

6. **The largest cities** in these States are Cincinnati, Louisville, and Cleveland, all admirably situated for trade. Cincinnati, like Chicago, has an extensive pork trade.

HISTORICAL GEOGRAPHY.

1. **KENTUCKY,** at first a part of the Territory of Virginia, was explored by Daniel Boone, just before the Revolution.

2. **OHIO and INDIANA** were first explored and claimed by the French.

3. **The first white inhabitants** of these three States were greatly annoyed by the Indians.

4. **Victories** over the Indians were won by Generals Wayne and Harrison.

COMPARATIVE SIZES.

Indiana is a little larger than Ireland.

Cincinnati, in 1870, had nearly the same population as Hamburg—216,000.

From Cincinnati as a centre, describe a circle of 100 miles radius; it will pass through Columbus and Indianapolis, and very near Louisville.

Observe that Ohio and Kansas, from north to south, are the same—200 miles; that Ohio and Indiana together are nearly the same as Kansas in extent from east to west—400 miles.

Observe that Kentucky is the same in length as Kansas, and almost the same in width.

Kansas, the common measure.

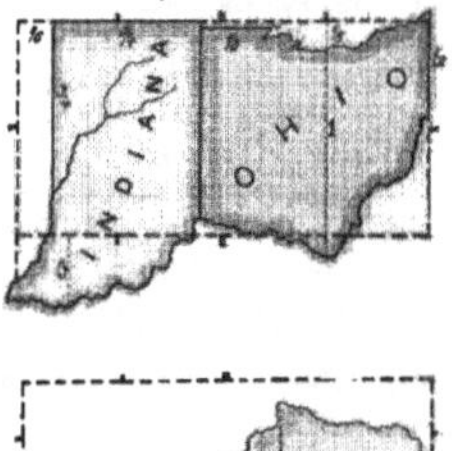

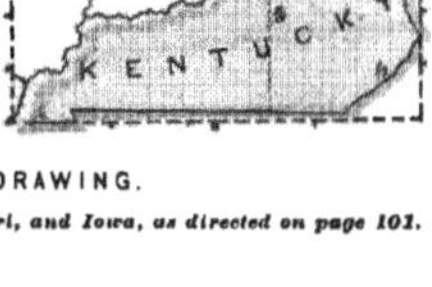

MAP-DRAWING.

Draw a Map of Illinois, Missouri, and Iowa, as directed on page 101.

Wheat and Cattle.

Minnehaha Falls, Minnesota.

Corn and Sheep.

REVIEW QUESTIONS

Mention the boundary rivers of Illinois, Missouri, and Iowa—the largest river in each of these three States—the capital of each State. Which of their cities are on the Mississippi River? Where are the lead mines? The iron mines? From St. Louis to Chicago, how many miles? How many hours by railroad? *Where are they?*

Rivers:—Illinois, Des Moines, Kaskaskia, Rock, Osage, Red Cedar, I'owa.

Cities:—St. Louis, Chicago, Kansas City, Peoria, Davenport, St. Joseph, Dubuque, SPRINGFIELD, Quincy, Bloomington, Burlington, DES MOINES.

PHYSICAL AND DESCRIPTIVE GEOGRAPHY.

1. **ILLINOIS** is remarkable for its fertile soil and its facilities for trade by lake, rivers, canals, and railroads.

2. **Its products** are abundant, consisting chiefly of grain, hay, live stock, lumber, coal, lead, and copper.

3. **Illinois excels** in the production of corn, wheat and oats.

4. **Navigation** between the Mississippi and the lakes is carried on by way of the Illinois River and a canal which connects it with Lake Michigan.

5. **The largest city in Illinois** is Chicago, which is remarkable for its rapid growth and its immense trade in grain, provisions, and lumber.

Among the advantages of its site are its fine harbor near the southern extremity of Lake Michigan, and its facilities for receiving and forwarding the products of a rich section of country near it.

6. **MISSOURI** has fertile prairie land in the north, and rich minerals in the south.

7. **The leading products** are corn, hemp, lumber, live stock, iron, lead, zinc, and copper.

8. **The largest city** is St. Louis, noted for commerce and manufactures; Kansas City and St. Joseph, on the Missouri River and important railroads, are the largest cities in the western part of the State.

9. **IOWA** consists mostly of rich prairie land.

10. **Its principal products** are grain, grass, live stock, wool, and lead.

11. **Railroads** to the Pacific pass through I'owa and Missouri.

12. **The largest cities** in Iowa are Davenport and Dubuque.

COMPARATIVE SIZES, ETC.

Missouri is larger than Maine, New Hampshire, Vermont, Massachusetts, and Rhode Island combined. Iowa is larger than England.

If you describe a circle of 175 miles radius around Terre Haute, it will pass near the three great cities of the West. Name them.

Observe that Iowa and Kansas are the same in width—200 miles, and that the length and breadth of Illinois are nearly the same as those of Kansas—200 by 400 miles.

HISTORICAL GEOGRAPHY.

1. **ILLINOIS** was settled by the French (in 1673), ceded to England (in 1763), and came into the possession of the United States government, at the Revolution.

2. **MISSOURI and IOWA** formed part of the Louisiana tract purchased from France by the United States (in 1803).

3. **The oldest town** in Missouri is St. Genevieve; in I'owa, Burlington. When Chicago was organized as a town, in 1833, it contained only 550 inhabitants. Now it is the fifth city, in size, in the Union.

TO DRAW WISCONSIN AND MINNESOTA.

Begin at F, and measure ½ m. north, and draw their southern boundaries, A B C D; thence 1⅓ north to H. Fix the points E, G, L, M, and N, as indicated, and complete the States.

TO DRAW MICHIGAN.

Form the square C N I F, each side 1½ m. long, and subdivide into four squares. Draw the ***Strait of Mackinaw, Lakes Michigan, Huron, St. Clair,*** and ***Erie;*** then the southern boundary. At S, ½ m. north of the ***Strait of Mackinaw,*** fix the southeastern extremity of ***Lake Superior;*** thence 1⅓ west to the western extremity, K, and draw ***Lake Superior,*** noticing that the northern coast at R is north of the west coast of ***Lake Michigan.***

REVIEW QUESTIONS.

Mention the boundary lakes—rivers. What great river rises in Minnesota? What rivers flow toward the north? Where is the copper region? Where are the lead mines?

A canal around the Falls of St. Mary, and another connecting the Wisconsin and Fox Rivers at Portage City, afford communication by steamboat between the Great Lakes and the Mississippi River.

How can you sail from St. Louis to Chicago? From Chicago to Cincinnati? From Chicago to St. Paul?

Where are they?

Lakes:—Superior, Michigan, Huron, Erie, St. Clair.

Rivers:—Minnesota, Kalamazoo, Wisconsin, Grand, Red River of the North, Saginaw, Detroit, St. Clair.

Cities:—Detroit, Milwaukee, ST. PAUL, Grand Rapids, Minneapolis, Oshkosh, Racine, MADISON, Fond du Lac, East Saginaw, Jackson, LANSING.

PHYSICAL AND DESCRIPTIVE GEOGRAPHY.

1. **MICHIGAN, WISCONSIN,** and **MINNESOTA** have vast forests in the north, and fertile prairie land in the south, and are rich in lumber and wheat; wool is also largely produced.

2. **Copper and iron** are abundant near Lake Superior.

3. **The commerce** of Michigan and Wisconsin is extensive, owing to the facilities afforded by the Great Lakes and the Mississippi River; and an important railroad will soon be built through Minnesota from Lake Superior to the Pacific.

4. **The Climate** of Minnesota is remarkable for its dryness.

5. **The largest city** in Michigan is Detroit; and in Wisconsin, Milwaukee, whose trade and manufactures are considerable.

6. **St. Paul** is at the head of navigation on the Mississippi.

7. **MICHIGAN** and **WISCONSIN**, like Illinois, were first held by the French, and afterward by the English.

8. **MICHIGAN** suffered much from incursions and massacres by the Indians. In the second war Detroit fell into the hands of the British (1812); but both the Indians and British were soon after defeated by General Harrison.

COMPARATIVE SIZES.

Michigan, Wisconsin, and Minnesota have an area almost equal to that of the German Empire.

They are in the same latitude as France.

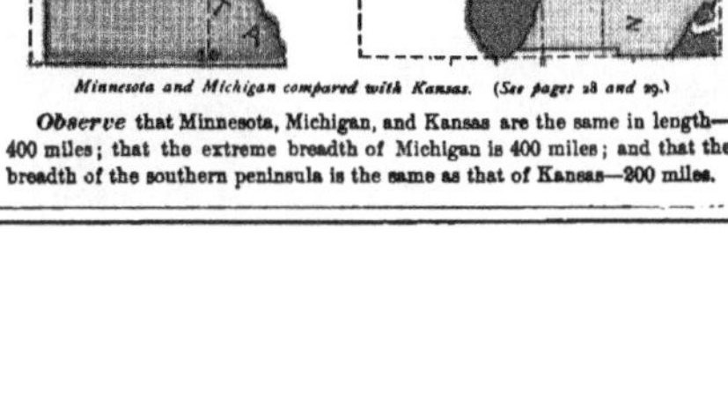

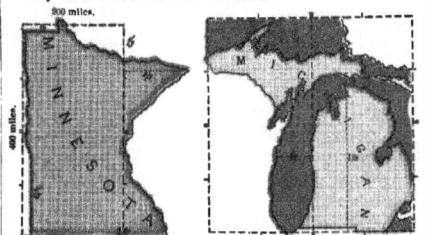

Minnesota and Michigan compared with Kansas. (*See pages 28 and 29.*)

Observe that Minnesota, Michigan, and Kansas are the same in length—400 miles; that the extreme breadth of Michigan is 400 miles; and that the breadth of the southern peninsula is the same as that of Kansas—200 miles.

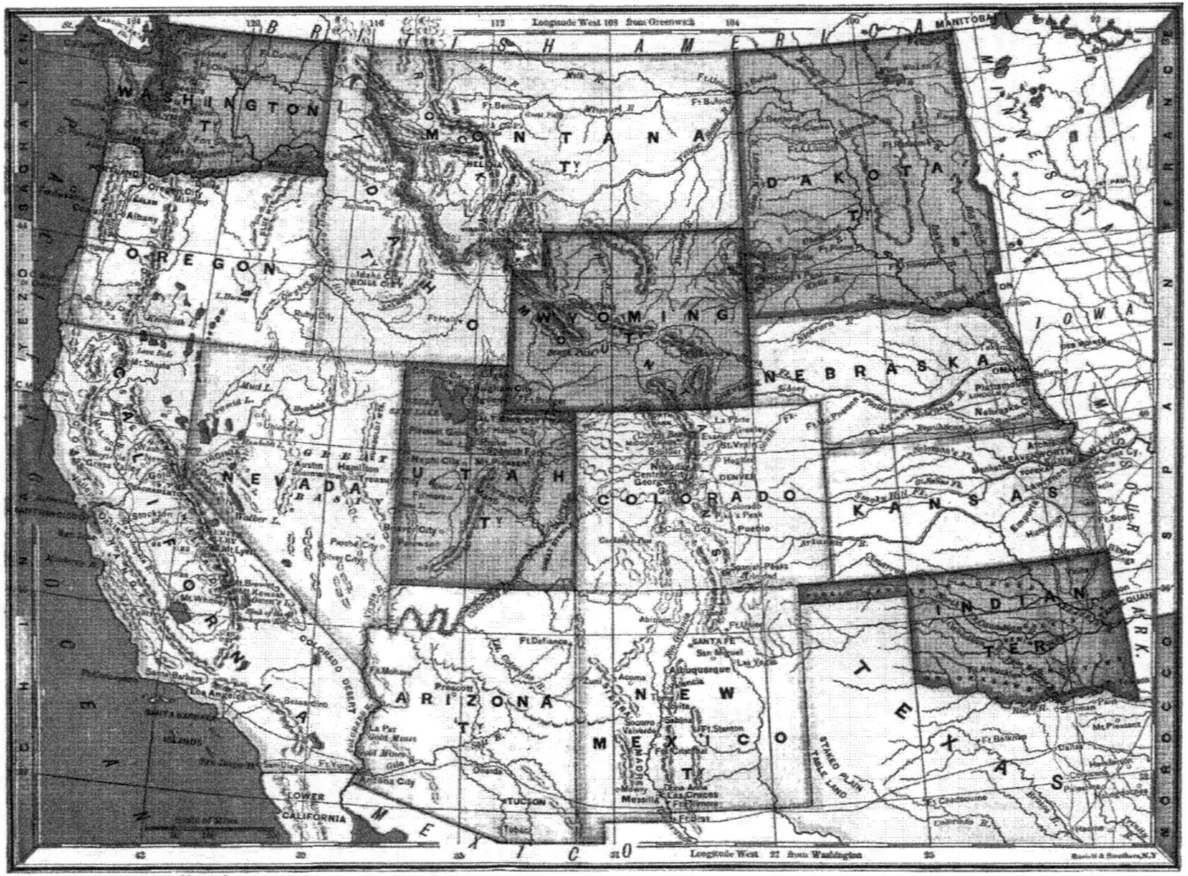

For the names of places represented by numbers. see p. 103

1. **THE WESTERN HALF OF THE UNION** is celebrated for high mountains, great mineral wealth, and barren table-lands.

2. **The Rocky Mountains** rest on the high table-lands which extend westward to the Sierra Nevada and the Cascade Range.

3. **The elevation** of the surface between these ranges is from 4,000 to 6,000 feet.

4. **A mild and delightful climate** is enjoyed in the valleys which lie west of the Sierra Nevada and the Cascade Range. (See p. 11, paragraphs 10, 11, and 12.) Rain is abundant there, while eastward the land is dry and barren.

5. **The Surface** slopes gradually from the Rocky Mountains eastward to the Mississippi River.

6. **The highest peaks** in the United States are about 15,000 feet above the level of the sea, nearly the same in elevation as Mt. Blanc, the most celebrated mountain in Europe. They are Mt. Whitney and Mt. Shasta, in California; Pike's Peak and Long's Peak, in Colorado; and Fremont's Peak, in Wyoming. Their summits are continually covered with snow.

GENERAL QUESTIONS.

What parallel of latitude forms the northern boundary of this section? What territories extend to that parallel? To what parallel does Maine extend? What States and Territory border on the Pacific Ocean? What Territories border on Mexico? On British America? What parts of Asia lie directly west? (*See margins.*)

Through what State and Territories do the Rocky Mts. extend?

What part of the Rocky Mountains are farthest from the Pacific? What high peaks in Colorado and Wyoming?

What large rivers have sources in Colorado? In Wyoming? In Montana? In Idaho? What rivers form boundaries? Name the branches of the Missouri,—of the Colorado,—of the Columbia,—of the Kansas River,—of the Arkansas River.

What river rises near Pike's Peak and flows into the Gulf of Mexico?

What rivers drain the Great Valley of California? What branch of the Columbia drains the eastern part of Oregon?

What States and Territories are bounded by the parallel of 42°? What parallel separates New York from Pennsylvania? What State and Territories have their southern boundaries on the parallel of 37°? What three Territories have their northern boundaries on that parallel? What four Territories are bounded by the meridian 32° west from Washington? How many degrees from Greenwich is that meridian? (*The difference between the meridian of Greenwich and that of Washington is 77°.*)

What bays on the Pacific coast? What capes? What strait?

TEXAS. (See p. 36.)

MAP-DRAWING.

Begin at its northwestern corner, and proceed east $\frac{3}{4}$ m.; thence south $\frac{3}{4}$ m. to the ***Red River;*** along ***Red River*** 1$\frac{1}{2}$ m. to ***Arkansas;*** thence $\frac{1}{2}$ m. south to the ***Sabine River,*** on the parallel of 32°; thence along said river, $\frac{3}{4}$ m. to its mouth.

Mark the most western point of ***Texas,*** on the parallel of 32°, 3$\frac{1}{2}$ m. across the State.

The most southern point is 1$\frac{1}{2}$ southwest of the mouth of ***Sabine River,*** and 3$\frac{1}{4}$ from the northeast corner of the State.

Complete the boundaries, and add the rivers and cities.

REVIEW QUESTIONS.

Where are they?

Rivers:—Rio Grande, Red, Sabine, Trinity, Nueces (*nway'ses*), Colorado, Brazos.

Cities and Towns:—Galveston, San Antonio, Houston, AUSTIN, Brownsville.

PHYSICAL AND DESCRIPTIVE GEOGRAPHY.

1. **TEXAS** is low along the coast, high and barren in the northwest. The interior contains rich prairie land. **The principal City** and port is Galveston.

2. **The people** are chiefly employed in farming and stock raising; cotton and corn are largely produced.

HISTORICAL GEOGRAPHY.

1. **TEXAS** belonged to Mexico until the revolution in 1836, when it became independent, upon the defeat of the Mexican forces and the capture of the Mexican president, Santa Anna.

2. **It contained** many emigrants from the United States, and was admitted into the Union in 1845.

3. **War** between the United States and Mexico was caused by disputes as to that portion of the state which lies between the Nueces and the Rio Grande.

4. **The Generals** were Scott and Taylor, of the Americans, and Santa Anna, of the Mexicans.

5. **All the battles were won** by the Americans, and the war ended when General Scott captured the city of Mexico.

6. **Mexico ceded** to the United States California and all the land eastward as far as the Rocky Mountains, for fifteen million dollars. (See page 101.)

NEBRASKA, KANSAS, AND INDIAN TERRITORY.

(TO BE DRAWN IN ONE MAP.)

Draw ***Kansas,*** an oblong, 1 m. wide by 2 ms. long; then the ***Kansas River,*** which empties into the ***Missouri River*** at the eastern boundary of the State; then the branches of the ***Kansas River,*** all emptying into it on the northern side.

In extent from north to south, ***Nebraska*** and ***Indian Territory*** are about the same as ***Kansas,*** 1 m. The eastern boundary of ***Indian Territory*** is $\frac{1}{2}$ m., and on a line with that of Kansas.

The western boundary of ***Nebraska*** is $\frac{1}{2}$ m. west of ***Kansas.***

Draw the rivers, and mark the cities and towns.

REVIEW QUESTIONS.

Where are they?

Rivers:—Missouri, Kansas, Arkansas, Platte, Keha Paha, Niobrara, Republican Fork, Canadian.

Cities:—Leavenworth, Omaha, Lawrence, Atchison, Nebraska City, TOPEKA, LINCOLN.

PHYSICAL AND DESCRIPTIVE GEOGRAPHY.

1. **The surface** of Kansas and Nebraska rises westward from the Missouri River.

2. **The soil** is well adapted to agriculture and stock-raising. The river valleys yield large crops of wheat and corn, and in most other parts grass is abundant.

3. **Timber** is scarce, but extensive beds of coal have been discovered in Nebraska.

4. **Kansas** lies in the center of the Union, being about 1,200 miles from either ocean.

5. **Its area** is greater than that of Denmark, Holland, Belgium, Switzerland, and Greece combined.

6. **The Union Pacific Railroad** crosses Nebraska, and the Kansas Pacific Railroad crosses Kansas.

7. **Omaha** is the eastern terminus of the Union Pacific Railroad, which runs westerly over the plains and mountains about 1,000 miles, to Great Salt Lake, where it connects with the Central Pacific Railroad.

HISTORICAL GEOGRAPHY.

1. **NEBRASKA and KANSAS** formed parts of the Louisiana tract purchased from France by the United States (in 1803) (See page 38, Historical Geography.)

COLORADO, UTAH, NEW MEXICO, AND ARIZONA.

MAP-DRAWING.

Observe that their opposite boundaries are parallel with each other, or nearly so; and that the 32d meridian and the 37th parallel form boundaries. On these two lines construct Colorado, Utah, New Mexico, and Arizona, by measuring from the point of intersection east, a little less than 2 ms., and north 1¼, to form ***Colorado;*** west, 1¼, gives the southwest corner of ***Utah;*** south, 2 ms., to the southeast corner of ***Arizona.***

Complete their boundaries; mark the mountain ranges and their principal peaks. Draw the rivers which have sources in Colorado, New Mexico, and tell where each empties; then the lakes and their inlets. Mark the capitals and principal towns.

REVIEW QUESTIONS.

Where are they?

Peaks:—Pike's, Long's, Fremont's, Spanish.

Rivers:—Colorado, Grand, Rio Grande, Gila, Green, Virgin.

Lakes:—Great Salt, Sevier, Preus, Utah.

Cities and Towns:—Salt Lake City, Denver, Golden City, Santa Fe, Ogden, Central City, Tucson.

PHYSICAL AND DESCRIPTIVE GEOGRAPHY.

1. **The High Section comprises** table-lands and valleys several thousand feet above the level of the ocean, while the mountains rise several thousand feet higher.

2. **Gold and silver** abound in all but Utah; mining is therefore the chief occupation.

3. **The lack of rain** renders large tracts of land unfit for cultivation; but the abundance of grass makes grazing profitable. Agriculture is carried on successfully in the river valleys, and timber abounds on the mountains.

4. **Utah is noted** for its lakes, which have no outlet to the ocean; and for the canyon through which the Colorado flows.

5. **The banks** of the Colorado are in some places more than 4,000 feet high, and very precipitous.

6. **The principal canyons** are between the Virgin River and the junction formed by the Green and Grand Rivers.

7. **Utah** is inhabited mostly by Mormons, who live in the vicinity of Great Salt Lake; their leading occupation is agriculture.

8. **Salt Lake City** is the largest city west of the Rocky Mountains, except San Francisco.

Quartz Mining.—Interior of a Tunnel.

CALIFORNIA AND NEVADA.

MAP-DRAWING.

On the 42d parallel draw the northern boundary of ***California,*** 1½ m., and of ***Nevada,*** 2 ms.

The southern point of ***Nevada*** is 2½ ms. south of the northeast corner of the State.

The western boundary of ***Nevada*** is 1 m. long, and parallel with its eastern boundary. Complete the boundary of ***Nevada,*** and draw the ***Colorado River*** to its mouth.

California is about 3½ ms. from north to south. Draw its southern boundary. ***San Francisco*** is 1½ m. from the northern boundary of the State, and ½ southwest from the angle formed by its eastern boundary lines.

Complete the boundaries of ***California,*** and mark the mountains, rivers, lakes, bays, capes, cities, and towns.

REVIEW QUESTIONS

Where are they?

Mountains:—Sierra Nevada, Coast Mts., Mt. Shasta, Humboldt, Whitney.

Rivers:—Sacramento, San Joaquin (*wah-keen'*), Humboldt, Klamath.

Lakes:—Pyramid, Tulare, Humboldt, Mud, Walker, Carson.

Cities and Towns:—San Francisco, SACRAMENTO, San Jose (*ho-say'*), CARSON, Stockton, Vallejo (*vahl-ya'ho*), Virginia City, Oakland, Marysville, San Diego, Grass Valley.

PHYSICAL AND DESCRIPTIVE GEOGRAPHY.

1. **The characteristics** of California are its mountain ranges with the great valley between them, its mineral and agricultural wealth, mild climate, high mountains, big trees, and scenery of surpassing grandeur and beauty.

A view of the Yo Semite Valley, in the Sierra Nevada, looking up the Valley (E. by N. E.). On the right or south, is the Bridal Veil Fall (630 feet); on the left, El Capitan, a perpendicular cliff (3300 feet).

2. **Gold** is found chiefly in the Sierra Nevadas; and quicksilver, in the Coast Range.

3. **The production** of wheat, wine, and wool is immense.

4. **The commerce** of California with China, Japan, and Australia is very important.

5. **The largest city** west of the Rocky Mountains is San Francisco. Sacramento, the capital, Oakland, Stockton, San Jose, and Marysville are important cities.

6. **NEVADA** is remarkable for its dry surface, great elevation, parallel ridges of mountains, and rich silver mines.

7. **It has access to the Pacific** by way of the Colorado River and the Gulf of California, the Colorado being navigable below the mouth of the Virgen River.

HISTORICAL GEOGRAPHY.

1. **CALIFORNIA** was discovered in the 16th century. It formed a portion of Mexico until 1848, the same year in which gold was discovered. It became a State in 1850.

2. **OREGON** was organized as a Territory in 1848, when it extended northward to British America and eastward to the Rocky Mountains.

3. **The eastern** parts of Colorado and New Mexico formed parts of the Louisiana purchase; the other parts, with the whole of Utah and Arizona, formerly belonged to Mexico.

Scenery on the Columbia River.—The Cascades.

OREGON, WASHINGTON, & IDAHO.

MAP-DRAWING.

On the parallel of 42°, draw the southern boundary of ***Oregon***, $1\frac{7}{8}$ m., and of ***Idaho***, $1\frac{1}{2}$ m. Draw the western boundary of ***Idaho***, 2 ms., the northern boundary, $\frac{1}{4}$ m., and complete ***Idaho***.

The northern boundary of ***Washington*** is $1\frac{1}{3}$ m. between ***Idaho*** and the ***G. of Georgia***. Mark ***Fort Walla Walla*** 1 m. south of ***British America***, and $\frac{1}{4}$ m. west of ***Idaho***; thence west $1\frac{1}{3}$ mark the mouth of the ***Columbia***, and draw that river, with its branches. Complete the boundaries of ***Oregon*** and ***Washington***, and mark their mountains, lakes, rivers, capes, cities, and towns; also ***Vancouver's Island***, the ***St. of Juan de Fuca***, and the ***G. of Georgia***.

View among the Sierra Nevadas.—Snow-sheds on the Pacific Railroad.—Donner Lake.

REVIEW QUESTIONS.

Where are they?

Mountains:—Cascade Range, Mt. St. Helen's.
Rivers:—Columbia, Clarke's, Lewis, Willa'mette.
Capes:—Flattery, Hancock, Foulweather.
Cities and Towns:—Portland, OLYMPIA, SALEM.

PHYSICAL AND DESCRIPTIVE GEOGRAPHY.

1. **The richest portions** of Oregon and Washington are between the Cascade Mountains and the Pacific Ocean; they contain fertile valleys and nearly all the population.

2. **The climate** of the western section is more mild and uniform than that of the Atlantic coast in the same latitude.

3. **Forests** are extensive, grass is abundant, and a flourishing trade is carried on in lumber, wool, and live stock.

4. **East of the Cascade Mountains** the land is high and dry.

5. **IDAHO** is rich in gold and silver.

WYOMING, DAKOTA, AND MONTANA.

MAP-DRAWING.

Draw a line running north and south $2\frac{1}{4}$ m. in length; one-half of this line will be the eastern boundary of ***Montana***, and the other half the eastern boundary of ***Wyoming***. Mark the width of ***Wyoming***, $1\frac{1}{4}$ m., and complete its boundaries, observing that the eastern boundaries of ***Colorado***, ***Wyoming***, and ***Montana*** are the same in length, $1\frac{1}{4}$ m.

Montana is a little more than $2\frac{1}{4}$ on its northern boundary, and its most southern point is $\frac{1}{4}$ m. south of the northwestern corner of ***Wyoming***. Complete the boundaries of ***Montana***.

The northern boundary of ***Dakota*** is $1\frac{1}{4}$ m. in length, and its southwestern corner is at the middle point of the eastern boundary line of ***Wyoming***.

Draw the mountains in these Territories, then the lakes, rivers, and towns, also the Yellowstone National Park.

REVIEW QUESTIONS.

Where are they?

Rivers:—Missouri, Yellowstone, Red R. of the North.
Cities:—Cheyenne, Helena, Yankton, Bannock.

PHYSICAL AND DESCRIPTIVE GEOGRAPHY.

1. **The characteristics of Montana** are its high mountains, mineral wealth, dry climate, and fertile valleys.

2. **Great natural wonders**, such as cañons, boiling and sulphur springs, geysers and volcanoes, abound in the Yellowstone National Park.

The Park is 55 by 65 miles in extent, and is under the control of the Secretary of the Interior.

3. **Helena** is the principal city in Montana.

MONTEITH'S RELIEF MAP of

m the PACIFIC OCEAN to the MISSISSIPPI RIVER.

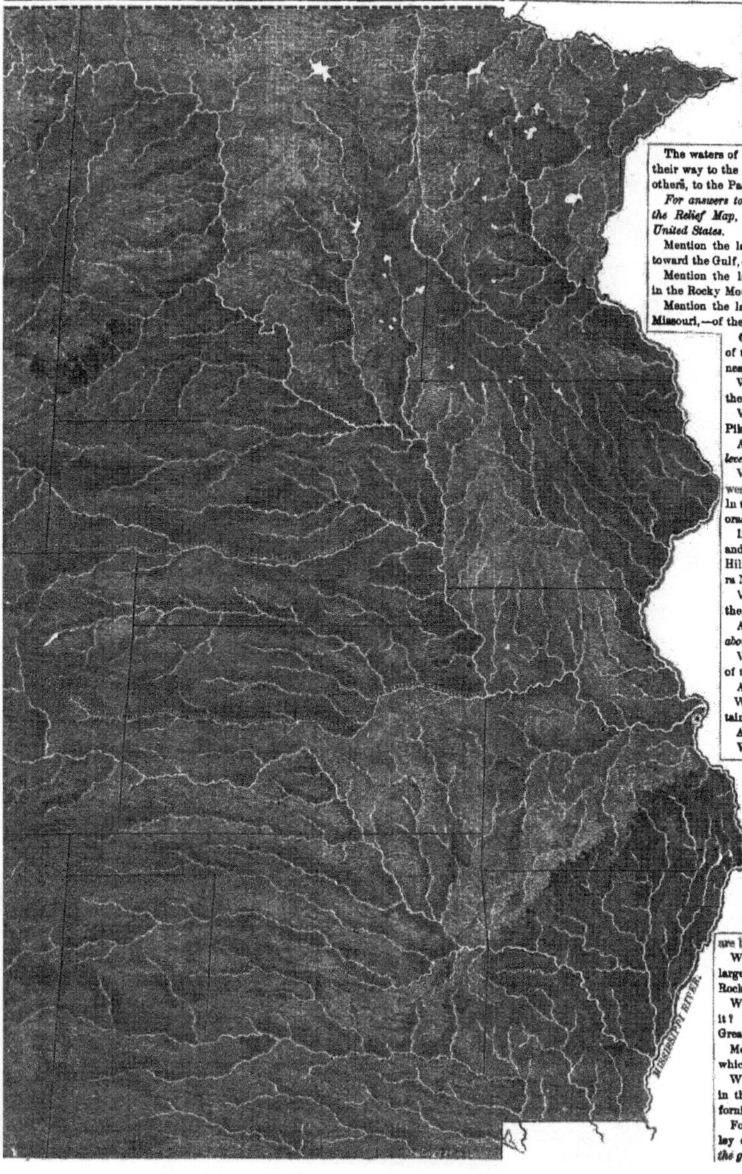

Observe that the Rocky Mountains are almost midway between the Pacific Coast and the Mississippi River, and that far up their sides are the sources of numerous streams and rivers.

The waters of some of these rivers find their way to the Gulf of Mexico; and of others, to the Pacific Ocean.

For answers to these questions, refer to the Relief Map, and to the Map of the United States.

Mention the largest rivers which flow toward the Gulf,—toward the Pacific.

Mention the largest river which rises in the Rocky Mountains.

Mention the largest tributaries of the Missouri,—of the Columbia.

Observe that the sources of these two rivers are very near each other.

What very high peak near the center of Colorado?

What is the height of Pike's Peak?

ANS. *14,500 feet above the level of the sea.*

What high peak in the western part of Wyoming? In the northern part of Colorado?

In what part of Dakota and Wyoming are the Black Hills? Where are the Sierra Nevadas?

What is the height of the Sierra Nevadas?

ANS. *About 15,000 feet above the level of the sea.*

What is the highest peak of these mountains?

ANS. ***Mount Whitney.***

What is the highest mountain in the United States?

ANS. ***Mount Whitney.***

Where is Mt. Whitney?

What is its elevation above the level of the sea?

ANS. *15,086 feet.*

Where is Mount Shasta? Mt. Hood?

Where is the Cascade Range? Mount St. Helens?

Which of the States represented on this Relief Map are best supplied with lakes?

What and where is the largest lake west of the Rocky Mountains?

What important city near it? What railroad passes Great Salt Lake?

Mention some of the lakes which have no outlets.

What small valley or gorge in the eastern part of California?

For what is Yosemite Valley celebrated? ANS. *For the grandeur of its scenery.*

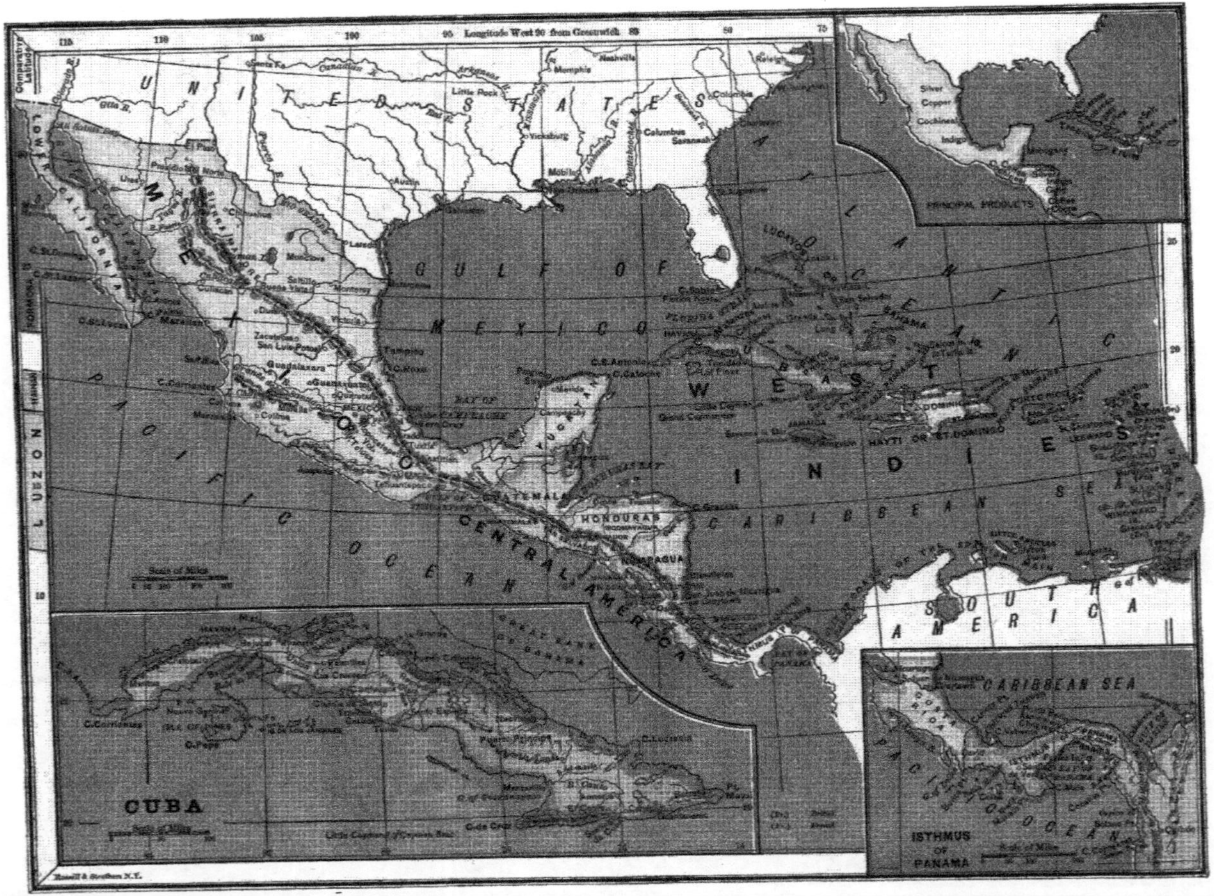
Longitude West 90 from Greenwich
UNITED STATES
MEXICO
LOWER CALIFORNIA
GULF OF MEXICO
ATLANTIC OCEAN
WEST INDIES
CARIBBEAN SEA
PACIFIC OCEAN
CENTRAL AMERICA
HONDURAS
NICARAGUA
GUATEMALA
YUCATAN
HAYTI OR ST. DOMINGO
SOUTH AMERICA
LUZON
FORMOSA
HAINAN
PRINCIPAL PRODUCTS
Silver
Copper
Cochineal
Indigo
CUBA
ISTHMUS OF PANAMA
CARIBBEAN SEA
Scale of Miles

MEXICO AND CENTRAL AMERICA.

1. **MEXICO** lies in two zones—the Torrid and the North Temperate.
2. **Its widest part** is in the north; and its narrowest, in the south.
3. **It extends** in a northwesterly direction.
4. **Its latitude** is the same as that of Arabia and the Great Desert.

GENERAL QUESTIONS.

What tropic passes over the middle of Mexico and near the northern coast of Cuba?
In what zone is the northern half of Mexico? The southern half?
Bound Mexico. What is its capital? Where is the nearest approach of the Gulf of Mexico to the Pacific Ocean?
What mountains in Mexico and Central America? What high volcano near the City of Mexico?
What gulf and peninsula in the west? What peninsula in the southeast?
Name the states of Central America. Bound Central America. What large lake in Nicaragua?

TO DRAW MEXICO AND CENTRAL AMERICA.

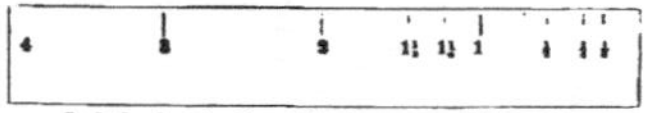

Scale for drawing the continents. 1 m. equals 600 miles.

Begin by marking a point as the centre of the Gulf of Mexico.

From this point, 1 m. eastward, mark ***Cape Sable;*** 1 m. westward, the coast of ***Mexico;*** ½ m. north, the mouth of the ***Mississippi River;*** and ½ m. south, the northwestern coast of ***Yucatan:*** then draw the whole coast of the ***Gulf of Mexico.***

What is the length of the Gulf of Mexico, in miles? How far is it from Yucatan to the mouth of the Mississippi River?

A view of the City of Mexico and vicinity.

Locate ***Mazatlan*** 1 m. west of the ***Gulf of Mexico.*** ½ m. west of Mazatlan mark the southwest coast of ***Lower California*** and locate ***Cape St. Lucas.***

A little more than 1 m. north of ***Mazatlan,*** locate ***El Paso,*** in the northeastern corner of ***Mexico,*** and the same distance west of ***El Paso,*** mark the northwest corner of ***Lower California.***

Draw the ***Rio Grande*** and locate ***Matamoras.***

Draw the northern boundary, and complete the coast of ***Mexico.***

2 ms. south of ***Cape Sable,*** mark the eastern extremity of ***Costa Rica,*** on the ***Mosquito Gulf;*** thence ¼ m. across to ***Cape Burica.*** How many miles? Complete the ***Pacific Coast.***

1½ a little west of south from ***Cape Sable,*** fix ***Cape Gracios,*** and complete ***Central America.***

TO DRAW THE WEST INDIES.

¼ m. south of ***Cape Sable,*** locate ***Matanzas*** and ***Havana,*** on the north coast of ***Cuba.*** How wide is ***Florida Strait?*** ¼ m. northeast from ***Cape Catoche,*** mark ***Cape St. Antonio,*** the western extremity of ***Cuba,*** from which to the eastern extremity is 1½ m. What is the length of ***Cuba?***

Hayti (*hay'te*) is one half the length of ***Cuba,*** and lies midway between ***Florida*** and the southeastern of the ***Little Antilles*** (*ahn-teel'*). How long is ***Hayti?*** Complete the map.

REVIEW QUESTIONS.

Where are they?

Gulfs and Bays:—California, Tehuantepec (*tay-won-tay-pek'*), Mosquito.

Capes:—St. Lucas, Palmo, Catoche (*kah-to-chay'*), Roxo, Gracios, Corrientes.

Cities:—Mexico, Guadalaxara (*gua-dah-lah-hah'rah*), Guatemala, San Salvador, Tampico, Guanaxuato (*gwah-nah-hwah'to*), Mazatlan, San Juan, Vera Cruz, Balize (*bah-leez'*), Comaya'gua.

Grand Plaza or Square in Guanaxuato, Mexico: showing Cathedral and government buildings.

PHYSICAL AND DESCRIPTIVE GEOGRAPHY.

1. **The characteristics of Mexico** are its great mountain chain, high plateaus, volcanoes, short rivers, silver mines, and tropical products.

2. **Its climate** and productions vary with the elevation. (See page 6, paragraph 28.)

3. **Along the coasts,** the climate is hot and pestilential; and on the high grounds, temperate and cold.

4. **The products** of the tropical regions are indigo, sugar, cotton, tobacco, oranges, etc.; and of the temperate regions, corn and wheat. Cochineal is an important article of export.

5. **The seasons** in the southern half of Mexico are two—the rainy (in summer), and the dry (in winter).

6. **The inhabitants** are Indians, Creoles, and mixed races.

NOTE.—The population of Mexico is 9,000,000, one-half ofwhom are Indians or native Mexicans. The Creoles are native whites of Spanish descent; they are the wealthy and influential class, and form one-sixth of the population. They inhabit the table-lands. They have a dark complexion, black hair and eyes. The Indians constitute the laboring class.

7. **The Government,** which is republican, is very unsettled, owing to revolutions.

8. **Education** of the masses has long been neglected, but is now receiving more attention.

9. **The better classes** are very gay; and all, male and female, are fond of smoking.

10. **The wild animals** of the hot region are the jaguar (*jag-u-ahr'*) and puma; and on the northern plains, roam immense numbers of bison, cattle, and horses.

11. **The City of Mexico,** the capital, is situated on a plateau more than 7,000 feet above the level of the sea. It is the largest city in the country, and has a population about equal to that of Cincinnati or New Orleans (200,000).

12. **Central America** resembles Mexico in its characteristics and inhabitants.

13. **It comprises** five republics—Guatemala, San Salvador, Honduras, Nicaragua, and Costa Rica.

HISTORICAL GEOGRAPHY.

1. **MEXICO was discovered** by Cordova, in 1517; and two years afterward it was conquered by Cortez, and remained in possession of the Spaniards about 300 years, or until it became a republic (in 1821).

2. **Many other Spaniards** afterward entered Mexico, and acquired great wealth from the gold and silver mines.

3. **The natives** were well advanced in the arts of civilization.

4. **Mexico** formerly extended north to Oregon, and east to Louisiana.

5. **Texas** declared itself free from Mexico, and was afterward admitted into the Union. War between the United States and Mexico soon followed. The Americans were commanded by General Scott, and the Mexicans by General Santa Anna; the former were victorious, and Mexico ceded to the United States, California, Nevada, Utah, and New Mexico, for $15,000,000.

6. **Mexico was invaded** by the French army in 1863, and Maximilian, of Austria, became emperor. He was soon taken and shot, and the republic was re-established.

THE WEST INDIES.

Name the largest of the West Indies. What group north of Cuba? East of the Caribbe'an Sea? In what zone is Cuba? What water between Cuba and Yucatan? Cuba and Hayti? Hayti and Porto Rico?

Into what two parts is Hayti divided?

What is the capital of Cuba? Of the Republic of Hayti? Of the Republic of Dominica? Of Jamaica?

Mention some of the Bahamas,—the Car'ibbee Islands.

Nassau, a town on New Providence Island, is the seat of government of the Bahamas. It has a good harbor and a healthful climate. It belongs to Great Britain.

PHYSICAL AND DESCRIPTIVE GEOGRAPHY.

1. **The WEST INDIES include** all the islands which extend from Florida, southeastwardly to South America.

2. **The largest** are Cuba, Hayti (*hay'te*), Jamaica, and Porto Rico (*meaning rich port*).

3. **The West Indies are celebrated** for their hot climate, destructive hurricanes and earthquakes.

NOTE.—Hurricanes are most frequent in August, September, and October. The Great Antilles (*ahn-teel'*) comprise Cuba, Hayti, Jamaica, and Porto Rico.

4. **The climate** is dry in winter; rainy and unhealthful in summer, when yellow fever is prevalent; snow is unknown.

5. **The productions,** which are abundant, include sugar, coffee, tobacco, cotton, corn, mahogany, dye-woods, drugs, pineapples, oranges, and bananas. The soil is cultivated by negroes.

6. **The inhabitants** are whites, negroes, and mulattoes, and the languages spoken, Spanish, French, and English.

7. **Fish,** turtles, parrots, reptiles, and insects are numerous.

8. **The most important city** is Havana, which is noted for its exportation of sugar, molasses, tobacco, and cigars.

9. **The population of Cuba** is equal to that of Massachusetts (about 1,400,000). The whites are of Spanish descent.

10. **To Great Britain** belong Jamaica, the Bahamas, Barbadoes, St. Vincent, Trinidad, and Antigua.

11. **To Spain** belong Cuba, Porto Rico, and the Isle of Pines.

12. **To France** belong Guadaloupe and Martinique.

13. **To Denmark** belong St. Thomas, St. John's, and Santa Cruz.

A circle of 700 miles radius drawn around New Orleans would pass near Havana, Tampico, Charleston, Cincinnati, and Kansas City.

Observe that the Island of Hayti is almost as long as Kansas, and that Cuba is twice the length of Hayti. (See page 98.)

HISTORICAL GEOGRAPHY.

1. **The WEST INDIES** were discovered in 1492. (See page 19, Lesson XI.)

2. **The natives** lived in bark huts, and obtained their food by hunting and fishing. They were nearly all killed or driven from the islands by the Spaniards.

3. **Slavery** has been abolished, except in the Spanish possessions.

4. **Cuba** has belonged to Spain ever since its discovery by Columbus.

5. **On the Island of Hayti or San Domingo** was founded the first Spanish colony in America.

6. **That Island now comprises** two republics, whose inhabitants are mostly negroes and mulattoes. The republic of Hayti occupies the western part, where the French language is spoken, and Dominica the remaining part, where Spanish is spoken.

TOPICAL GEOGRAPHY.

NOTE.—These exercises may be used orally or made the subjects for composition; and, if the teacher wish, he may require the pupils to draw the State on the blackboard, and, in turn, to mark the places when they are mentioned. These questions may also be applied to any other State.

The State you live in:—Between what parallels of latitude is it? What are its boundaries? What can you say of its size? Its shape? Surface? Productions? Manufactures? Rivers? Principal cities? History?

NORTH AMERICA.—In what zones? Mention its countries—capes—inlets—lakes—principal rivers. What can you say of its mountains—surface—climates—rain—dry regions—frozen regions—agricultural products—mining—fisheries? Which coast is best adapted to commerce? Why? Mention its largest cities. Which are noted for commerce? Manufactures? What parts of North America are noted for cotton—oranges and lemons—gold and silver? What State is especially rich in grain, gold, and quicksilver? In silver? What States are noted for copper? Lead? Coal and iron? Lumber? Corn and wheat? What State excels in shipbuilding? Commerce? What Territories abound in gold and silver? What two cities are noted for their trade in grain and pork?

What can you say of the people of the different countries—their color, habits, dress, language, and occupations? Of the government and history of each country?

What animals are in the northern regions? The temperate? The tropical?

What and why is the difference between the climates of the Pacific coast and those of the Atlantic coast of the United States and British America? What State is celebrated for great trees? What is the principal seaport on the Atlantic coast? On the Pacific coast? What are some of the advantages of their situation? How does New York compare in population with London? With Berlin?

Why is the Pacific coast of the United States so well watered? Why is the land east of the Sierra Nevada dry and barren?

From what States and islands do we get sugar? What city is noted for its export of sugar and cigars? Of gold? Of cotton?

What part of Mexico is healthy? Unhealthy? Why? Where are its cities—on the highland or lowland? What season is the best in which to visit the West Indies? Why?

What parts of North America belong to Great Britain? To Denmark? What islands belong to Spain?

Where are seals, turtles, cod, and mackerel caught? What do you know of them and their uses?

What advantages for trade have Chicago and Buffalo?

What three Atlantic States have no mountains? What Gulf States have no mountains?

SOUTH AMERICA
VENEZUELA
U.S. OF COLOMBIA
ECUADOR
PERU
BRAZIL
BOLIVIA
ARGENTINE REPUBLIC OR LA PLATA
PARAGUAY
URUGUAY
CHILI
PATAGONIA
Georgetown
Paramaribo
Cayenne
Cuzco
Trinidad
Cochabamba
Sucre
Potosi
Port Alegre
San Paulo
ASSUMPTION
Cordova
BUENOS AYRES
MONTEVIDEO
PRINCIPAL PRODUCTS.
Indigo
Chocolate
India Rubber
Cocoa Nut
Peruvian Bark
Nitre
Silver
Coffee
Sugar
Cotton
Diamonds
Gold
Paraguay Tea
Wool
Hides
Tallow
S. Limit of Wheat
COMPARATIVE AREA
IOWA
ILLINOIS
MISSOURI
BORNEO
AUSTRALIA
CAPE COLONY
GUINEA
Longitude West from Greenwich

SOUTH AMERICA.

1. **SOUTH AMERICA lies** chiefly in the Torrid Zone. Its southern part is in the South Temperate.

2. **Its area** is equal to three-fourths that of North America.

3. **Its eastern point** is due south of Cape Farewell.

4. **Its western point** is directly south of Cape Sable (Florida), and its northern cape is west of the central point of Africa.

5. **Its shape** is triangular; its widest part being from Cape St. Roque to the northwestern part of Peru.

GENERAL QUESTIONS

What countries are crossed by the Equator? By the Tropic of Capricorn? What countries are wholly within the Torrid Zone? Within the South Temperate Zone?

What is the direction of the coast from its northern to its eastern cape? From its eastern to its southern cape?

By what oceans and sea is South America surrounded?

What mountains extend along the Pacific coast? The Atlantic coast? What mountains south of Venezuela? In the western part of Brazil? What volcanoes in Ecuador? Between Chili and Argentine Republic?

Into what ocean do nearly all the rivers flow? Name the largest rivers.

Name the largest three tributaries of the Amazon? What large tributary has the Parana? What country is embraced between the Parana and Paraguay rivers? What large rivers of Brazil flow into the Atlantic Ocean?

Name the northern cape—the eastern—the southern.

What lake west of Bolivia? In Venezuela?

What strait between Patagonia and Terra del Fuego?

What countries border on the Atlantic? On the Pacific? On the Caribbe'an Sea? What country has no sea-coast?

To what three nations does Guiana belong? Bound Colombia,—Venezuela,—Guiana,—Brazil,—Ecuador,—Peru,—Bolivia,—Paraguay,—Argentine Republic,—Uruguay,—Chili,—Patagonia. What is the capital of each?

On a voyage from the West Indies to Rio Janeiro, what countries and prominent capes would you pass? In what directions would you sail on a voyage from Rio Janeiro to Valparaiso? From Valparaiso to Panama?

What large island at the mouth of the Amazon?

Give the situation of the principal seaports of South America,—Rio Janeiro,—Bahia,—La Guayra,—Valparaiso,—Callao,—Guayaquil,—Panama—Maranhao

DIRECTIONS FOR DRAWING SOUTH AMERICA.

[See "General Directions" on Page 17.]

Commence at A, and measure 5½ ms. north, and mark *Cape St. Roque.* Measure to C, 7½ ms.; thence west to D, 5½ ms. At 4 ms. from C, mark *Cape Gallinas* and *Lake Maracaybo.* From 5½, at *Cape St. Roque,* toward 4 on the line C D, mark the points, 1, near the mouth of the *Amazon River;* 2, opposite *Georgetown;* and, 3, near *Caracas.* Complete the coast line.

From A, toward the *west,* mark the points 4 and 5½ at B. From 4, west of A, toward *Cape St. Roque,* mark 1, near the *Gulf of St. George;* 2, opposite *St. Matthias' Bay;* 3, near the mouth of the *Rio de La Plata;* and 6, opposite the *Bay of All Saints.* Complete the coast line.

From 4, west of A, measure 4 ms. north, and draw the coast south to *Terra del Fuego* and *Cape Horn.*

North of B, mark the points 5½, 6, 7, and draw the *Gulf of Darien, Isthmus of Panama,* and *Cape Blanco.* Complete the drawing by marking the mountains, rivers, countries, bays, gulfs, capes, cities, etc., writing the *full* name of each outside the map.

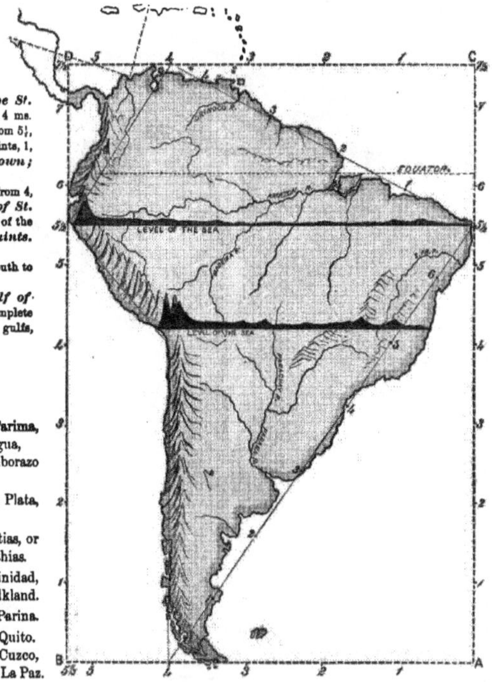

REVIEW QUESTIONS.

Where are they? (See Note, p. 15.)

Mountains:—Andes, Pacaraima, Parime, or Parima, Geral, Vol. Co'topaxi, Vol. Aconcagua, Vol. Pichincha (*pe-cheen'chah*), Mt. Chimborazo (*rah'zo*).

Rivers:—Amazon, Orinoco, Uruguay, Negro, La Plata, Paraguay, Parana, St. Francisco.

Gulfs and Bays:—Darien, Panama, San Matias, or Venezuela, St. George, St. Matthias.

Islands:—Terra del Fuego, Marajo, or Joannes, Trinidad, Chiloe (*che-lo-ay'*), Chonos Arch., Falkland.

Capes:—Gallinas (*gal-le'-nas*), St. Roque, Horn, Parina.

Cities:—Rio Janeiro, Buenos Ayres, Valparaiso, Quito. Bahia (*bah-e'a*), Lima, Santiago, Cuzco, Montevideo, Bogota', Pernambuco, La Paz.

PHYSICAL AND DESCRIPTIVE GEOGRAPHY.

1. **South America is celebrated** for its great mountain chain, immense plains and rivers, tropical climate, and the abundance of its vegetable and animal life.

2. **Its three mountain systems** are in the west, east, and north. (See Relief Map.) The greatest is that of the Andes, which extend along the Pacific coast.

3. **The Andes** are from 8,000 to more than 20,000 feet elevation. The highest peak is Aconcagua, nearly 24,000 feet.

NOTE.—The highest mountains in North America are about 18,000 feet; in Asia, 29,000 feet; Mt. Blanc is 15,800 feet.

4. **The highest peaks of the Andes** are continually covered with snow.

View among the Andes.—Travelers crossing the mountains.

5. **The base of the Andes** is about 400 miles wide.

6. **In the center and north,** the Andes comprise two or more ranges, between which are plateaus containing lakes and cities. The plateau of Bolivia is more than two miles above the level of the ocean.

What lake and cities are on this plateau?

7. **The other mountains** and table-lands are in Brazil, Guiana, and Venezuela; but they are not so high as those of the Andes.

8. **Nearly all** else of South America is a vast plain, drained by the Amazon, the Orinoco, and the La Plata rivers.

9. **The plain of the Amazon** is noted for its dense forests, called silvas, which are the abode of savages, monkeys, alligators, serpents, and insects, besides birds of wonderful beauty and variety.

10. **In the forests** are trees from which cocoa, India-rubber, Peruvian or quinine bark, cabinet and dye-woods are obtained.

11. **The lowlands** of the Orinoco, called llanos (*l'yah'noce*), and those of the La Plata, called pampas, are covered with grass in the wet season, from November to May; but in the dry season, from May to November, they resemble a desert.

12. **The llanos and pampas** afford pasture to vast numbers of cattle, horses, and sheep. The wild cattle are caught by means of the lasso, a long leathern rope.

13. **The chief exports** from Uruguay and the Argentine Republic are tallow, hides, horns, hair, wool, and dried beef; from Bolivia and Peru, silver, guano, and nitre.

14. **Brazil produces** more than half the coffee used in the world.

15. **The principal productions** of nearly all the countries are coffee, sugar, cotton, tobacco, cocoa, and tropical fruits.

16. **Among the animals** are the jaguar (*jag-u-ahr'*) or American tiger, tapir, puma, ant-eater, sloth, alpaca, and armadillo. Mules and llamas are used as beasts of burden. The condor, the largest bird of flight, has its home in the Andes.

17. **The precious and useful metals** are found in the Andes; diamonds and other precious stones, in Brazil.

18. **The inhabitants of South America** comprise whites, Indians, negroes, and mixed races. The whites are in a small minority. In Colombia and Venezuela, only one-fourth are white.

19. **The white inhabitants** of Brazil are of Portuguese descent, and of the other countries, Spanish; in Brazil they live near the coast, while in the interior some of the native tribes are cannibals.

20. **Brazil** is an empire.

21. **Brazil is nearly as large** as the United States; and Uruguay, the smallest country in South America, is larger than the six New England States combined.

22. **In the Torrid Zone** is all that part of South America which lies north of the Argentine Republic, or about three-fourths of the whole division.

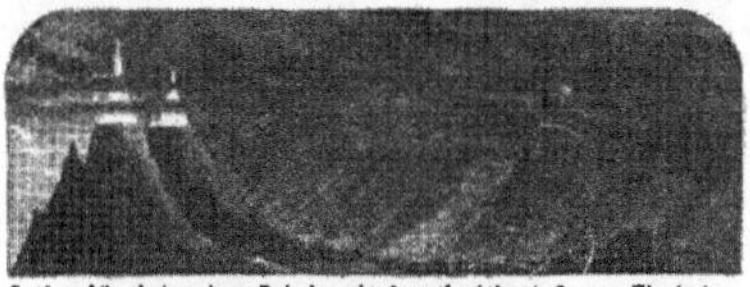

Section of South America.—Rain brought from the Atlantic Ocean.—The Andes.—The Great Plains.

23. **In the tropical countries** of South America, the winds blow from the Atlantic Ocean, and supply the vast plains with abundant rain. The moisture carried by these winds is condensed before passing the snow-covered peaks of the Andes; hence the rainless districts between the Andes and the Pacific.

HISTORICAL GEOGRAPHY.

1. **SOUTH AMERICA** was discovered by Columbus, at the mouth of the Orinoco (in 1498).

2. **Brazil was settled** by Portuguese, in the early part of the 16th century. It declared itself free from Portugal (in 1822) and became an empire. The Portuguese language is spoken there.

3. **The Pacific Ocean was discovered** by Balboa, who crossed the Isthmus of Panama (in 1510); and soon after, by Magellan, who entered the Pacific through the strait which bears his name, and thus made the first voyage around the globe.

4. **Balboa** was followed by Pizarro, a cruel Spaniard who conquered Peru, which then included nearly the whole of the western part of South America and was inhabited by a powerful and civilized race.

Natural bridges in Colombia, 300 feet above the torrent at the bottom.

NOTE.—The Peruvians—native Indians—had been governed by the Incas for centuries. They worshiped the sun, and believed their rulers, the Incas, to be its descendants. Their knowledge of architecture and sculpture is shown by the remains of palaces, aqueducts, and temples. The magnificent "Temple of the Sun" was richly ornamented with gold and jewels.

5. **Peru was held by Spain** for about three hundred years, and in the early part of the present century all the Spanish colonies, from Venezuela to Chili, inclusive, became independent.

6. **Guiana** belongs partly to Great Britain, France, and Holland; and Patagonia is claimed by Chili and the Argentine Republic. All the other countries are republics, except Brazil.

7. **The South American republics**, like Mexico, have suffered much from revolutions.

NOTE.—The Republic of Colombia, which continued for several years previous to 1831, comprised New Granada, Ecuador, and Venezuela, now three separate republics. New Granada is now known as the United States of Colombia.

TOPICAL GEOGRAPHY.

(TO BE USED AS A REVIEW OR AS SUBJECTS FOR COMPOSITION.)

SOUTH AMERICA:—What can you say of its position on the globe? Its shape and measurements? Its mountain ranges, volcanoes, and plateaus? Its climates? Rains and drouths, and their effects? Vegetable products? Minerals? Exports? Animals? Inhabitants, original and present? Governments? Discovery and history?

If the winds of the tropical countries blew from the Pacific instead of the Atlantic, would Brazil have abundant rains, large rivers, and dense forests? Why? Would it, in that case, be wet or dry on the west side of the Andes? Where, then, would be the rainless districts?

If the Andes were along the Atlantic coast, instead of the Pacific, what would be the effect? In Patagonia and southern Chili, the winds blow from the Pacific; on which side of the Andes, in those countries, does the most rain fall? Where are their rainless districts?

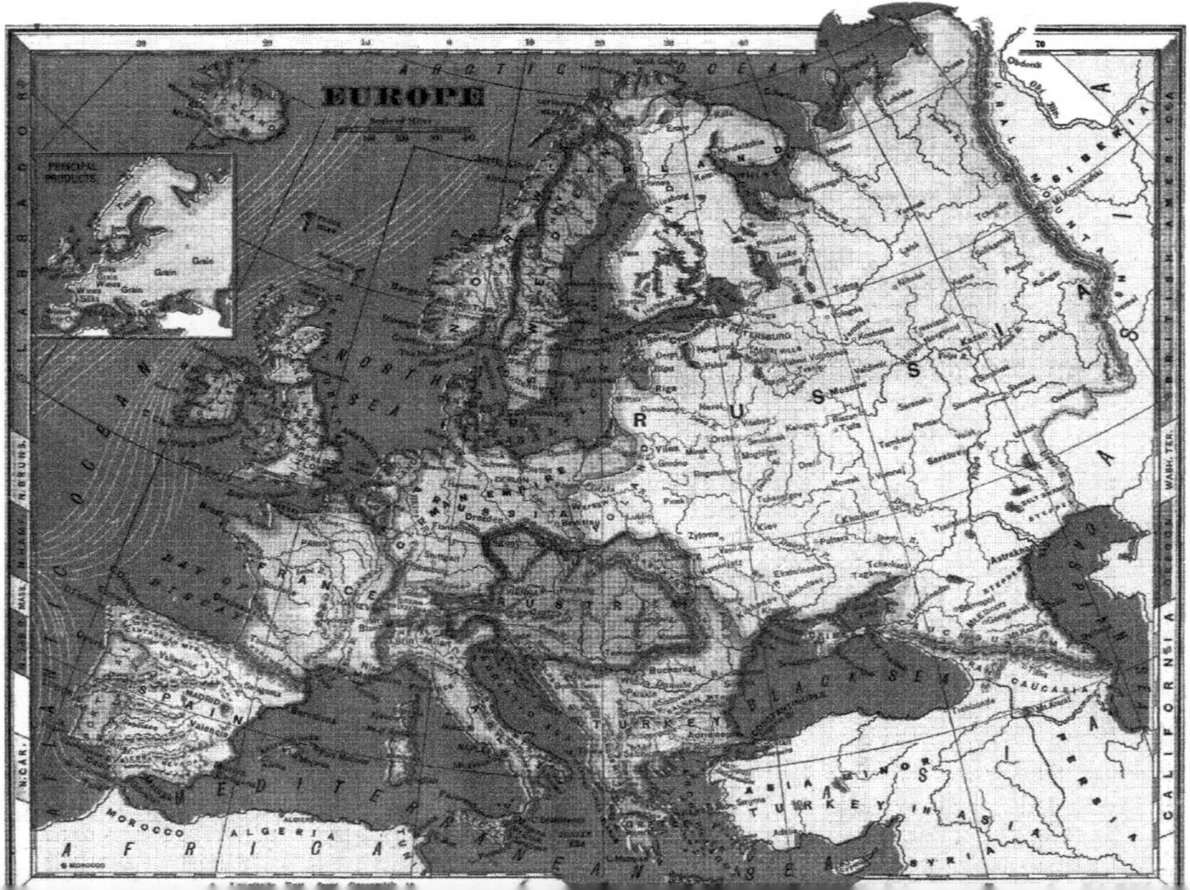

EUROPE
PRINCIPAL PRODUCTS
ARCTIC OCEAN
NORTH SEA
ATLANTIC OCEAN
BAY OF BISCAY
MEDITERRANEAN SEA
BLACK SEA
CASPIAN SEA
RUSSIA
SWEDEN
NORWAY
FRANCE
SPAIN
AUSTRIA
PRUSSIA
GERMAN EMPIRE
TURKEY
ASIA MINOR
TURKEY IN ASIA
PERSIA
SYRIA
CAUCASIA
SIBERIA
URAL MOUNTAINS
ASIA
AFRICA
MOROCCO
ALGERIA
LABRADOR
BRITISH AMERICA
CALIFORNIA
OREGON
WASH. TER.
N. BRUNS.
N. CAR.
MASS.

EUROPE.

1. **EUROPE** is situated in the northern half of the North Temperate Zone, west of the northern half of Asia, and east of the northern half of North America.

2. **Its extent from east to west** is about the same as that of the United States (2,800 miles).

3. **If Europe were moved westward,** and placed over the United States, so that Cape Matapan would be over Raleigh, the Strait of Gibraltar would be over Santa Fé, and Paris, on the most northern boundary of the United States.

4. **North Cape,** the most northern point of Europe, is in the same latitude as the most northern point of Alaska, in North America; and the City of St. Petersburg is in the same latitude as Cape Farewell.

GENERAL QUESTIONS.

Between what parallels of latitude is Europe? Through what countries of Europe does the parallel of 50° pass? Through what country and two large islands of North America does that parallel pass? *Ans. Canada, Newfoundland, and Vancouver's Island.*

What ocean north of Europe? What sea opens into it? What peninsula is formed by the White Sea?

What ocean west of Europe? What bay and seas open into the Atlantic Ocean? What two countries in the northwestern part of Europe form a peninsula? What two in the southwest form a peninsula? What islands west of the North Sea?

What sea south of Europe? What seas open into it? What peninsulas and capes project into it? What islands in it? By what is it connected with the Atlantic Ocean? With the Black Sea?

What peninsula north of the Black Sea? By what waters is it formed? On what waters and past what capes and islands would you sail on a voyage from the Crim-e'a to Sicily? From Sicily to Denmark? From London to St. Petersburg? From London around Great Britain?

What is the largest country in Europe? By what waters is Russia bounded? By what countries? What is the shortest route by water from Russia to Italy? Russia to England?

Has Switzerland any sea-coast? What mountains between Switzerland and Italy? What four large rivers of Europe rise among the Alps?

In what direction and through what countries does the Dan'ube flow? The Rhine? The Rhone? The Po? What rivers flow into the North Sea? Into the Baltic Sea? Into the Black Sea? Into the Caspian Sea? Into the White Sea? Mention the largest lakes in Europe.

What mountains in Austria? In Turkey? In Norway? In Spain? In Italy? East of Russia? South of Russia? Between France and Spain?

In what direction do the Apennines extend? The Alps? The Cau'casus? The Ural?

In what direction do the rivers of Prussia flow? Of Sweden? Of France? Of Spain?

From what mountains and hills does the Volga receive its water?

Name the countries of Europe. What is the capital of Russia? Of Turkey? Italy? Spain? Portugal? France? Austria? Switzerland? Greece?

DIRECTIONS FOR DRAWING EUROPE

Begin at A, and mark the points B, C, and D, and the points between them, as indicated. Next mark the points between E and F, H and C, and draw in order the ***Straits of Gibraltar,*** the eastern part of the ***Mediterranean Sea, Italy,*** the ***Adriatic Sea,*** the ***Archipelago,*** the ***Sea of Marmora,*** the ***Black Sea,*** and the islands ***Corsica, Sardinia, Sicily, Candia,*** and ***Cyprus.*** Next draw the ***Caspian Sea.***

Mark the points P and O, and the points between E, C, and E, K. Draw in order the coasts of ***Spain, France,*** and ***Holland,*** the ***Baltic*** and ***North Seas.*** Complete the northern coast line.

Mark the lines L O, and M P, and mark the coast of ***Great Britain, Ireland,*** and ***Iceland.*** Draw the mountains and rivers, locate the principal cities, and complete the map.

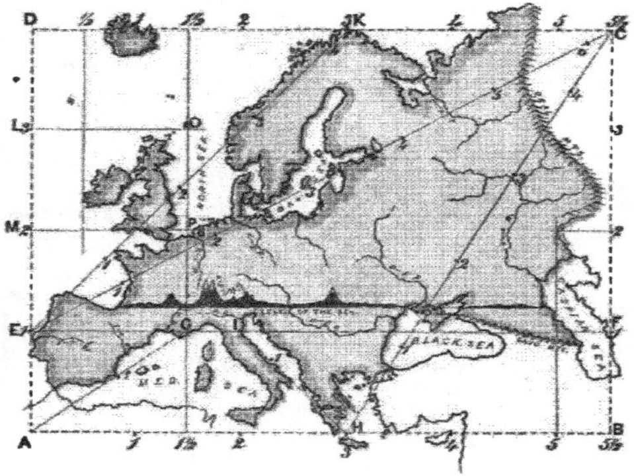

REVIEW QUESTIONS.

Where are they? (See Note, page 15.)

Mountains:—The Alps, Ural, Apennines, Pyrenees, Carpathian, Balkan, Caucasus, Scandinavian, Cantabrian, Dovrefield.

Rivers:—Volga, Dan'ube, Ural, Vist'ula, Dnieper, Don, Duna, Dwina, Seine, Loire, Rhone, Rhine, Po, Elbe, Ebro, Guadiana, Save, Tagus, Oder.

Seas:—North, Baltic, White, Irish, Caspian, Black, Az'ov, Adriatic, Archipelago, Marmora, Mediterranean.

Gulfs and Bays:—Bothnia, Onega, Finland, Lyons,* Biscay.

Straits, etc.:—Gibraltar, Bos'porus, Dardanelles, Yenikale (*yen-e-kah'-lay*), Dover, Cattegat, Skager Rack, English Chan., St. George's Chan.

Islands:—Sicily, Sardinia, Bal-e-ar'ic, Crete or Candia, Shetland Is., Corsica, Hebrides, Orkney Is., Gothland, Majorca, Ionian Is.

Capes:—North, Matapan, Finisterre* (*tare'*), Clear, Ortegal, St. Vincent, Land's End, The Naze,* Spartivento.

Cities:—London, Paris, Constantinople, Vienna, Berlin, Madrid, Lisbon, Dublin, St. Petersburg, Brussels, Bern, Rome, Edinburgh, Naples, Copenhagen, The Hague, Florence, Frankfort, Stockholm.

* G. of Lyons, G. of Lions, or G. of the Lion. Finisterre, Lat. for land's end. *The Naze*, nose.

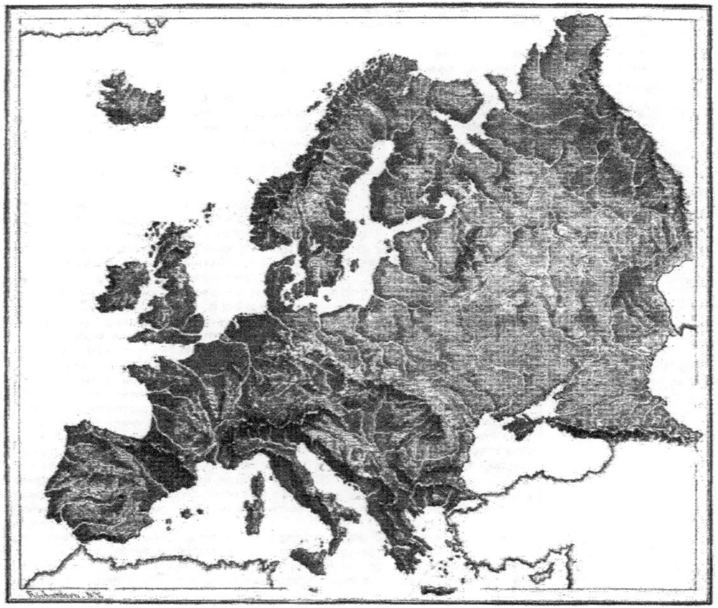

Relief Map of Europe.

EUROPE.

1. **EUROPE** is remarkable for the number and importance of its inlets, islands, and peninsulas, and consequently for its great length of coast line, which facilitates trade and navigation.

2. **Its northern half** is mostly level; its southern half, mountainous.

3. **Its principal mountains** are the Alps, Pyrenees, Carpathians, and the Dovrefield and Kiolen Mountains of Norway. Point to each on the Relief Map.

4. **The great plain** of Europe extends over Russia, Denmark, Holland, and the northern part of Prussia.

5. **The rivers of Russia** have their sources in the Ural Mountains and the Valdai Hills; those of Sweden, in the Dovrefield and Kiolen Mountains; while the largest rivers of the southwestern half of Europe rise in the Alps or the Carpathian Mountains.

6. **The highest mountain** in Europe is Mt. Blanc, whose summit is perpetually covered with snow. (Height, 15,810 feet.) Among the other celebrated peaks of the Alps are Mt. Rosa, the Jungfrau (*yoong'frow*), Mt. Cenis (*seh-ne'*), Mt. Cervin or the Matterhorn, Mt. St. Bernard, and Mt. St. Gothard (*got'hart*).

7. **The rivers** which are celebrated for the beauty of their scenery are the Rhine, Seine, Loire, and Rhone.

8. **The largest lakes** are in Sweden and Western Russia; the most picturesque are in Switzerland, Italy, and Scotland.

9. **The latitude** of Europe corresponds with that of British America and the northern part of the United States.

10. **The climate** of Western Europe is greatly modified by the westerly winds which blow over the warm waters of the Gulf Stream. (See chart, p. 89: also p. 11, paragraphs 10 and 11.)

11. **The heat of Southern Europe** is sometimes greatly increased by the hot winds from Africa.

12. **In the warm countries** of the south the winters are short, frost and snow are rare; and among their products are grapes, oranges, lemons, figs, citrons, and olives.

13. **Central Europe** has a temperate climate, with cold winters; it produces wheat in abundance.

Questions on the Relief Map.—In what part of Europe is the great mountain region? What mountains are on the boundary of Europe? In what countries are the largest lakes? In what direction does the surface of Sweden slope? Of Spain? The northern part of Prussia? Western France? Southern Russia? Northern Russia? Western Russia? In what directions does the surface of Italy slope? Where is the largest valley of Italy? What river drains it? By what mountains is it inclosed? What plain is drained by the Danube? What small plain in Austria is inclosed by mountains and drained by the Elbe?

THE COUNTRIES OF EUROPE.

COUNTRIES.	CAPITALS.	COUNTRIES.	CAPITALS.
ENGLAND, } (K)	*London.*	BELGIUM, (K)	*Brussels.*
SCOTLAND, } (K)	*Edinburgh.*	HOLLAND, (K)	*The Hague (haig).*
IRELAND, } (K)	*Dublin.*	SWITZERLAND, (R)	*Bern.*
FRANCE, (R)	*Paris.*	ITALY, (K)	*Rome.*
SPAIN, (K)	*Madrid'.*	BAVARIA, (K)	*Munich.*
PORTUGAL, (K)	*Lisbon.*	WURTEMBERG, (K)	*Stuttgart.*
THE GERMAN EMPIRE,	*Berlin.*	NORWAY, SWEDEN, } (K)	*Stockholm.*
PRUSSIA, (K)	*Berlin.*	RUSSIA, (E)	*St. Petersburg.*
AUSTRIA, (E)	*Vienna.*	TURKEY, (E)	*Constantinople.*
DENMARK, (K)	*Copenhagen.*	GREECE, (K)	*Ath'ens.*

NOTE.—(K) *signifies Kingdom*; (E), *Empire*; (R), *Republic.*

Map Drawing.—From the most northern point of England, southerly on the 2d meridian, to St. Alban's Head, the middle of the southern coast, is 1¾ m. (each m. = 200 miles). The southern coast from Land's End to North Foreland is 1⅗ m.

General Questions.—Bound England,—Wales. What rivers empty into the North Sea? Into the Irish Sea? What city on the Mersey River? On the Thames? Where is the Severn River? What capes or heads on the southern coast? Eastern? Western? What islands near Land's End? What English islands near the coast of France? Where is the Isle of Wight? What two large islands in the Irish Sea? What strait separates Anglesea from Wales? Where is Bristol Channel? Solway Firth? The Wash? Cardigan Bay?

Windsor Castle, the principal residence of the English monarchs. It is 23 miles S.W. from London. The buildings and court cover more than 12 acres of ground.

PHYSICAL AND DESCRIPTIVE GEOGRAPHY.

1. **ENGLAND is diversified** by hills and plains, on which are fine pastures and some of the best cultivated farms in Europe.

2. **The mountains and lakes** are chiefly in the northwest.

3. **Its highest peaks** are Scawfell, Helvellyn, and Skiddaw, each about 3,000 feet elevation, or one-half the height of Mt. Washington.

4. **Its largest lake** is Windermere, which is only ten miles long and one wide.

5. **The climate** is mild, moist, and healthful; rain is abundant, and large crops of wheat, barley, oats, and hay are produced.

England is farther north than Canada. What and why is the difference between their climates? (See pages 11 and 19.)

6. **England possesses** coal and iron in abundance, excellent harbors, and an industrious population; it, therefore, has large cities, with extensive manufactures and commerce, in which it excels every other country in the world.

7. **The principal manufactures** are cotton and woolen goods, hardware, and cutlery.

8. Celebrated for the manufacture of ***Cotton goods***, is Manchester; ***Woolen goods*** and ***Cloths***, Leeds; ***Iron and Steel ware***, Birmingham; ***Cutlery***, Sheffield; ***Silks***, Norwich; ***Ribbons***, Coventry; ***Carpets***, Kidderminster; ***Hosiery***, Nottingham. (How is each of these cities situated?) Population of England, 21,500,000; of Wales, 1,200,000.

Newcastle-on-Tyne, noted for its coal trade, which has been carried on 600 years. It received its name from the castle whose square tower is seen on the right of the picture.

9. **The population**, wealth, and commerce of London are greater than those of any other city in the world.

10. **Liverpool** is noted for its extensive commerce.

11. **Coal** is obtained chiefly in the north (in Durham and Northumberland counties, especially near Newcastle); **Iron**, from Wales, Stafford, and Yorkshire; **Lead**, from the northwest; **Tin** and **Copper**, from the southwest.

12. **The English and Welsh** are mostly stout, and have a ruddy complexion.

13. **The characteristics of Wales** are its hills, mountains, valleys, and fine pastures.

14. **Flannel** is extensively manufactured.

COMPARATIVE SIZES.

15. **England and Wales** together have the same area as Georgia. The population of London (over three millions) is nearly twice that of Paris, over three times that of New York, or four times that of Berlin.

16. **England and Florida** are the same in length.

HISTORICAL GEOGRAPHY.

1. **ENGLAND** was first called Albion, or white island, which referred to the chalk cliffs along the southern shore.

2. **It was invaded** by the Romans, under Julius Cæsar, about 50 years B. C. He gave it the name of Britannia (painted), because the inhabitants, who were Celts, stained their bodies.

3. The Celts were an ancient race of people, who formerly inhabited Central and Western Europe. Their descendants now inhabit Ireland, Wales, the Highlands of Scotland, and the northern shore of France.

4. **England was held** by the Romans 400 years, when it was abandoned by them. It was afterward conquered by German tribes called Angles and Saxons; hence the names Angle-land (England) and Anglo-Saxon.

5. **In the ninth century** Egbert and Alfred-the-Great distinguished themselves as rulers of England.

6. **The country was frequently invaded** by Danes and Norwegians; the Danish king Canute gained possession in the 11th century; but all were soon after conquered by the Normans from France.

7. **In the Norman line of succession** were the sovereigns William I, William II, Henry I, and Henry II. The latter became king in the 12th century.

8. **England was long disturbed** by contests with France and Scotland, and between rivals for the throne.

9. **The American Revolution** occurred during the reign of George III, grandfather of Queen Victoria.

10. **The Government** is a limited monarchy, and the crown is hereditary (descending from an ancestor).

11. **Parliament**, composed of the House of Lords and House of Commons, corresponds to our Congress.

12. **The United Kingdom** of Great Britain and Ireland comprises the British Isles, and with vast possessions in America, Asia, Africa, and Oceanica, forms the British Empire.

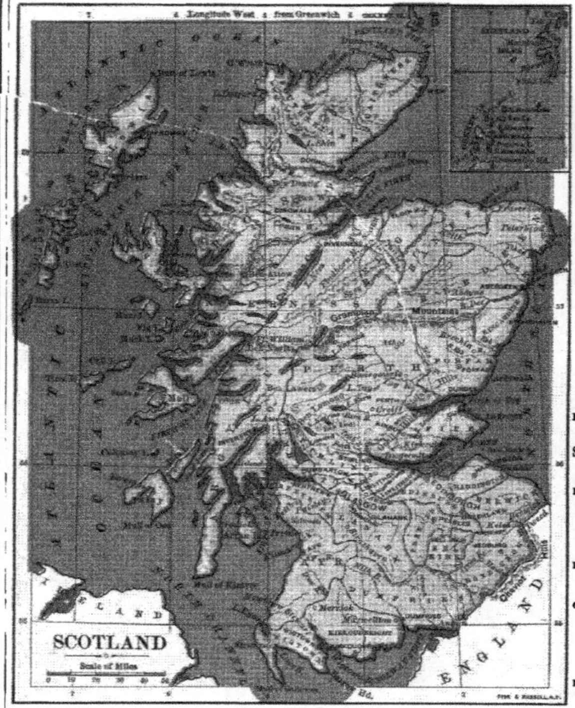

GENERAL QUESTIONS.

What meridians run through Scotland? What parallels? What is the most northwestern cape? The most southern? The most northeastern cape or head? What bay or firth south? What two large firths east? West? What large islands west and northwest of Scotland? What mountains in Scotland? What hills between Scotland and England? What channel between Scotland and Ireland? Where is Loch Lo'mond? The Clyde River? Dee River? Tay River? Edinburgh? Glasgow? Inverness? Perth? Aberdeen? Population of Scotland, 3,359,000.

View in the Highlands.—Highlander and the favorite instrument, the bagpipe.—Shepherd boy and girl.

Scotland in the frame of Kansas.

Observe that the widest part of Scotland, including the island of Skye, is 200 miles, and from north to south it is about 300 miles, or ⅜ the length of Kansas.

Edinburgh (*ed'in-bur-rah*), the capital, is a handsome city, and is noted for its institutions of learning.

Glasgow excels every other city in Scotland in population, manufactures and commerce.

PHYSICAL AND DESCRIPTIVE GEOGRAPHY.

1. **The characteristics** of Scotland are its broken coast line, rugged surface, moist climate, numerous lakes and islands.

2. **The Highlands** cover the northern, and the Lowlands the southern half of Scotland.

3. **The Highlands** are remarkable for wild and beautiful scenery—lochs, crags, ravines, and ruined castles. The Highlanders are chiefly shepherds.

Ben Nevis is the highest mountain, and Loch Lomond the largest lake, in Great Britain.

4. **The Lowlands** are comparatively level and fertile, and more thickly settled.

5. **The manufactures** include cotton, woolen, and silk goods. Nearly all the material of which cotton goods are manufactured is supplied by the United States.

6. **The people** are tall, hardy, thoughtful, and industrious, and many are highly educated. The Lowlanders are Anglo-Saxons; the Highlanders, Celts.

HISTORICAL GEOGRAPHY.

1. **The original inhabitants** were tribes of shepherds and hunters. Their religion was druidical, a form of idolatry.

2. **The Lowlands** were held by the Romans from A. D. 80 to 446, when the Picts became the ruling race. These were overpowered by the Scots from Ireland, whose kings held the country for several centuries.

3. **In the 13th and 14th centuries** England made great efforts to subjugate Scotland, but its independence was secured after a long war, in which Robert Bruce distinguished himself.

4. **Scotland and England** were united under one crown in 1603.

5. **Christianity** was introduced by missionaries from Ireland, in the 6th century.

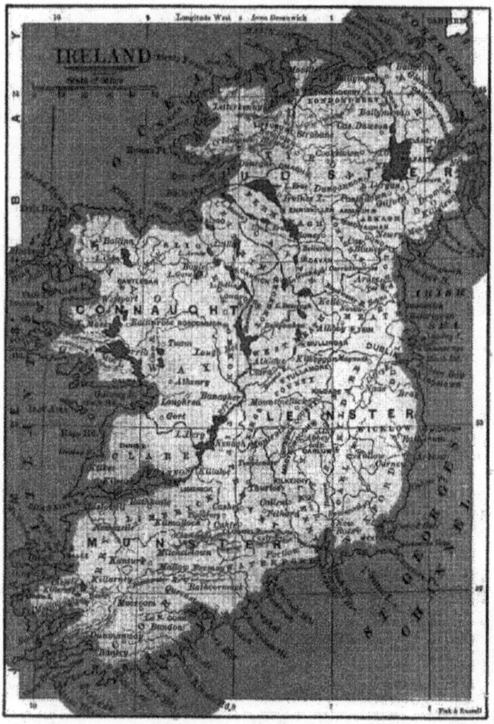

PHYSICAL AND DESCRIPTIVE GEOGRAPHY.

1. **IRELAND** is generally level, with mountains along the coast.

2. **Its climate** is mild and moist, and the people are chiefly engaged in grazing and agriculture.

3. **The principal products** are grain, potatoes, hay, and flax; live stock and linen are largely exported to England and Scotland.

4. **Its rivers and lakes** are numerous; Lough Neagh (*loh-nay*) is the largest lake in Ireland, and the Lakes of Killarney are celebrated for their beautiful scenery.

5. **The principal cities** are Dublin, Belfast, and Cork; Dublin is one of the finest cities in Europe.

Dublin is in the same latitude as the southern point of Alaska. Dublin and Cork together have a population about equal to that of Boston.

Ireland, with one-ninth the area of the six Gulf States, has a larger population (5,402,000).

6. **Bog-peat** is extensively used for fuel; coal, which is imported from Great Britain, is used only in the towns.

HISTORICAL GEOGRAPHY.

1. **The name** of Ireland was originally Ir or Erin; it was called Hibernia by the Romans. It is sometimes called the Emerald Isle, because of the freshness of its verdure.

2. **Ireland** was long governed by its own kings, princes, or chieftains; its four provinces, Ulster, Leinster, Munster, and Connaught, were former kingdoms.

3. **From the 3d to the 10th century** it was ruled by a tribe called the Scots, who called it Scotia, but afterwards transferred the name to Scotland. (See page 63, paragraph 2 of Historical Geography.)

4. **Invasions and insurrections** were frequent, and the country finally fell into the hands of the English.

5. **The people** belong chiefly to the Celtic race; they are active, industrious, warm-hearted, and impulsive.

GENERAL QUESTIONS.

What parallels of latitude pass over Ireland? Over what parts of North America do those parallels pass? (See p. 16.)

In what direction is Ireland from the United States? (See page 89.)

What ocean north and west of Ireland? What sea east? What channel southeast? Northeast?

What is its northern head or cape? Southern?

What city and bay in the middle of the eastern coast? What city and bay on the western side nearly west of Dublin?

Which is the most northern of its four provinces? Southern? Eastern? Western?

What lough or lake in the northeastern part of Ireland? What small lakes in the southwestern part?

In what part of Ireland is Belfast (celebrated for its linen and cotton factories? Cork (manufactures, ship-building, and fine harbor)? Limerick? Waterford? Londonderry? Galway? Donegal? Sligo?

Observe that Ireland is about the same in width as Kansas—200 miles, and that Kansas has a greater area than Ireland, Sardinia, and the peninsula of Denmark, combined. Ireland, Portugal, and the State of Indiana are about the same in area.

View in Ireland.—Peasantry and city gentry.—An Irish jaunting-car.

FRANCE. What parallels of latitude pass over France in this map? Over what parts of North America do these parallels pass? (See p. 16.) What ocean west of France? What sea south-east? What bay west? What channel northwest? What strait north? What country is separated from France by the Strait of Dover?

What countries border on the northeast? East? What mountains east? Mention the four large rivers of France. Into what does each empty?

What large island in the Mediterranean Sea belongs to France? *Ans. Corsica.*

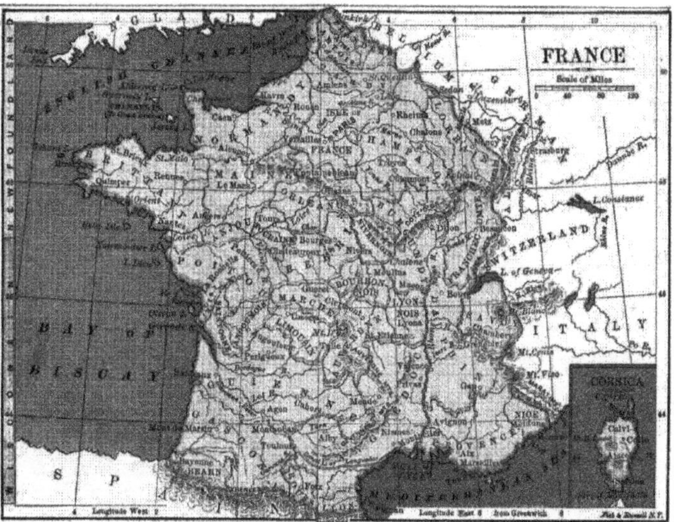

1. **FRANCE** is mostly level, with high mountains on the east and south.

In the east are Mt. Blanc, the highest mountain in Europe, and Mt. Cenis (*seh-ne'*), through which is a tunnel nearly eight miles in length, connecting France and Italy.

2. **Its climate,** except in the northeast, is mild. Its southeastern coast is a favorite winter resort for invalids from Europe and the United States.

3. **The leading agricultural products** are grain and grapes, besides olives, oranges, and lemons in the south. Around Lyons, the second city, in size, in France, mulberry trees are extensively cultivated for the support of the silk-worm.

4. **Among the exports** are silk, woolen, linen, and cotton goods, wine, brandy, jewelry, and fancy articles.

5. **Paris** is, next to London and Pekin, the largest city in the world. The most important seaport in France is Marseilles.

6. **The French** are remarkable for politeness, activity, and honesty; they are gay, impulsive, and fond of military distinction.

7. **France,** although smaller than Texas, has a population almost as large as that of the United States.

Population of France, 36,000,000; United States, 38,925,000.

8. **Its foreign possessions** comprise Algeria, French Guiana, besides parts of India, Western Africa, and several islands in Oceanica and the West Indies.

HISTORICAL GEOGRAPHY.

1. **FRANCE**, with Holland, Belgium, and Western Germany, was formerly called Gaul; the inhabitants were called Gauls or Celts.

2. **It was conquered** by Julius Cæsar and annexed to the Roman Empire (51 B. C.).

3. **It received its name** from the Franks, German invaders, who afterwards governed it.

4. **Charlemagne** was the greatest monarch of his time (8th century). By his conquests, his empire included France, Belgium, Holland, Germany, and Switzerland, with parts of Italy and Spain.

5. **The most celebrated** general and monarch was Napoleon Bonaparte, who became emperor of France and king of Italy (1804). He was overthrown at Waterloo (1815), and died in banishment in St. Helena. His nephew, Louis Napoleon, became president of France in 1848, and soon afterwards, emperor.

6. **Louis Napoleon,** in the war with Germany, in 1870 and 1871, lost his throne; and France, that part of its territory which bordered on the Rhine. (France lost Alsace and a part of Lorraine.)

A village in the eastern part of France.—The Vosges Mountains.

SPAIN AND PORTUGAL form the Iberian Peninsula. By what bodies of water is it surrounded? What countries join it on the northeast? What mountains between France and Spain?

What is the direction of the northern boundary of Spain? The eastern? What two capes on the northwest? What three on the coast of the Mediterranean? What is the southwestern cape of Portugal? What mountains in the north?

What four large rivers flow west and southwest? What is the largest river which flows into the Mediterranean Sea?

Mention the principal cities on the Douro,—On the Tagus,—on the Guadiana,—on the Guadalquiver,—on the Ebro,—on the eastern coast,—on the southern coast.

What islands east of Spain?

Madrid, the capital and metropolis of Spain, is in the same latitude as New York, Naples, and Salt Lake City? Where is Madrid?

PHYSICAL AND DESCRIPTIVE GEOGRAPHY.

1. **The surface** is a table-land, on which are five chains of mountains, extending from east to west; between these chains run the principal rivers.

2. **The interior** is high, dry, and treeless; its summers are hot and its winters cold; while in Portugal rain is abundant, and along the coast of the Mediterranean perpetual spring prevails.

3. **The chief exports** are wine, olive oil, wool, quicksilver, oranges, lemons, raisins, almonds, corks, and licorice.

4. **The useful metals** are found in the north; silver and quicksilver, in the south.

5. **The Spaniards and Portuguese** have dark eyes and are of a dark complexion. The Spaniards are proud and revengeful, and very fond of music, dancing, bull-fights, and other sports.

6. **The inhabitants** are descended from the Iberians and Celts, intermingled with Carthagenian, Roman, Gothic, and Moorish blood. (See Ancient Geography, page 90.)

7. **The Spanish and Portuguese languages**, although differing from each other, are both derived from the Latin.

8. **Education and agriculture** are in a very backward condition; scarcely one-tenth of the inhabitants of Spain being able to read and write.

9. **The most populous provinces** are in the south.

10. **Andorra** is a small republic in the Pyrenees, inhabited by shepherds. (Area, 149 square miles; population, 12,000.)

Population of Spain, 16,500,000; of Portugal, 4,000,000.

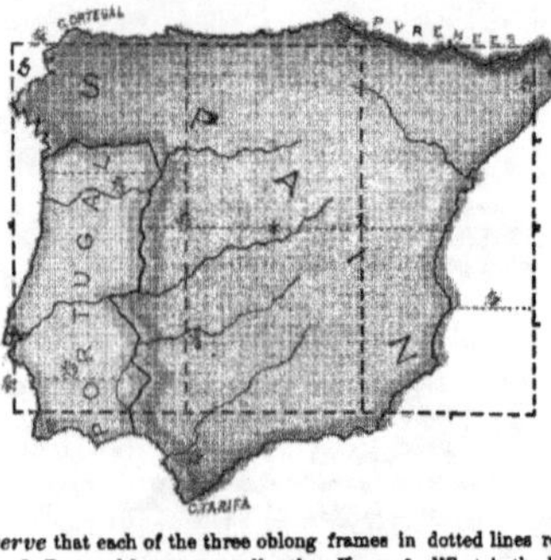

Observe that each of the three oblong frames in dotted lines represents Kansas. Is Portugal larger or smaller than Kansas? What is the length of the northern boundary of Spain?

HISTORICAL GEOGRAPHY.

1. **The peninsula was originally peopled** by the Iberians, a fierce race. On its coasts, colonies were established by the Phœnicians about 1000 B. C. It was subdued by the Carthagenians, and afterwards by the Romans, who held it at the beginning of the Christian era.

2. **At the fall** of the Roman Empire, tribes from Germany and France took possession; but these were succeeded by the Moors, or Arabs from Northern Africa (in the 8th century). The Moors were driven out by the Spaniards, under Ferdinand and Isabella, king and queen of Spain (1492), the same year in which Spain came into possession of large portions of North and South America, by right of discovery. Spain became a republic in 1873.

3. **The Portuguese**, in the 15th and 16th centuries, were among the most enlightened and enterprising people of Europe; their colonies along the coasts of South America, Africa, and Asia were very important. (See page 57, paragraph 2.)

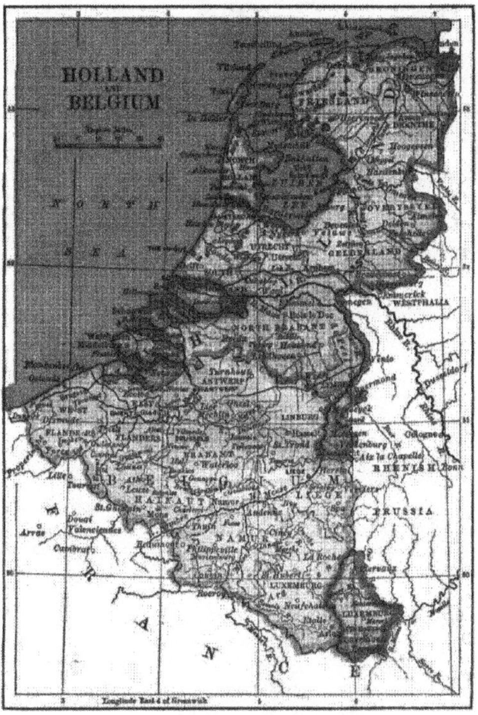

Holland.—Skating scene.

COMPARATIVE GEOGRAPHY.

Belgium, with one-fourth the area of New York State, has a larger population.

Zuider Zee (meaning *South Sea*) is in the same latitude as James' Bay.

PHYSICAL AND DESCRIPTIVE GEOGRAPHY.

1. **The words Holland and Netherlands** signify a flat, low, marshy country. A considerable portion of the surface of Holland and Belgium is more than twenty feet below tide water. Inundation is prevented by dikes or mounds.

2. **The low land is drained** by means of its numerous canals, which are between high banks, and into which the water is raised by windmills and steam-engines.

3. **The leading occupations** in Holland are commerce and dairy-farming.

4. **The climate** is moist and disagreeable.

5. **Belgium** is noted for its dense population, whose employments are chiefly in agriculture and manufactures.

6. **The Dutch and Belgians** are noted for their industry and cleanliness. Low-Dutch is spoken in Holland; Flemish (resembling Dutch) and French, in Belgium.

7. Population of Belgium, 5,000,000; Holland, 3,900,000.

HISTORICAL GEOGRAPHY.

1. **Holland and Belgium**, at the beginning of the Christian era, were subject to Rome. In the 8th century, they formed part of the dominions of Charlemagne; and afterwards, of Austria and Spain.

2. **Holland became independent** and very prosperous in the 17th century, and her foreign possessions comprised Sumatra, Java, and several other Asiatic islands, besides Dutch Guiana.

3. **Both countries** were afterward incorporated with France; but in 1815, they were joined together as the Kingdom of the Netherlands.

4. **They have been separate kingdoms** since 1831.

The oblong frame shows the length and breadth of Kansas (200×400 *miles*).

HOLLAND, OR THE NETHERLANDS, is ½ m. wide by 1 m. long (100×200 miles). By what is it bounded on the north and west? On the east? South?

What sea or zee in Holland? What is the largest river flowing through Holland? Does the Rhine empty through one or several mouths? What grand duchy southeast of Belgium belongs to Holland? (A grand duchy is governed by a grand duke.) What branch of the Rhine bounds Luxemburg on the southeast? Are there any mountains in Holland or Belgium? Bound Belgium. What rivers flow through Belgium?

Amsterdam, the metropolis, is one of the most important commercial cities in Europe. It is built on wooden piles driven into swampy ground. Where is it situated? Rotterdam, the second city of Holland, is, like Amsterdam, intersected by numerous canals. Where is it? The Hague (*meadow*) is the capital. Where is it?

Brussels, the capital of Belgium, is celebrated for lace and carpets, printing, publishing, and trade. Where is it? Ghent is the second city in Belgium, and is celebrated for its cotton manufactures. Where is it? Antwerp (meaning *added*, because built on successive deposits from the river) commands the commerce of Belgium. Where is it situated?

Alsace and that portion of Lorraine here shown, were ceded by France to Germany, in 1871.

THE GERMAN EMPIRE. What sea north of Germany? Northwest? What two gulfs open into the Baltic Sea? What countries bound Germany on the west and northwest? On the south? On the east? What four large rivers flow through Germany? In what direction do they flow? Into what do they empty? In what part is Rhenish Prussia? Westphalia? Pomerania? Schleswig and Holstein? Hanover? Prussian Saxony? Silesia? Brandenburg?

In what direction does the surface of Prussia slope?

Where and on what river is Berlin (the capital and largest city)? Hamburg (the chief commercial city in Germany)? Breslau (celebrated for its linen trade)? Dresden (works of art)? Cologne (cologne-water)? Frankfort (inland trade and banking business)? Bremen (2d commercial city)? Posen? Hanover?

Population of German Empire, 41,000,000.

PHYSICAL AND DESCRIPTIVE GEOGRAPHY.

1. **THE GERMAN EMPIRE** comprises Prussia, Bavaria, Wurtemberg, Baden, Saxony, and other states, besides free cities.

2. **Prussia** is mostly low and level, with mountains in the south. The winters on its northern plain are severe.

3. **The mineral products** include coal, salt, and the useful metals.

4. **Grain** and grapes are extensively cultivated throughout Germany.

5. **Merino sheep,** valuable for their wool, are raised in great numbers.

6. **The forests** are extensive.

7. **Prussia excels** every other country in its national system of education. Attendance at school between the ages of five and fifteen is required by law. Schools, universities, academies, and public libraries are numerous throughout Germany.

8. **The kingdoms in Germany** are Prussia, Bavaria, Wurtemberg, and Saxony.

9. **The Germans** are thrifty, industrious, of a calm temper, and fond of music, books, and study.

10. **Parents** pay great attention to the discipline and amusement of their children, and the love of home is a national characteristic.

A Vineyard in Germany.—Gathering Grapes.

A view of the Rhine,—Ruins of an ancient castle.

10. **The Rhine** is one of the most important rivers in Europe for navigation and trade.

11. **It is celebrated** for the beauty of its scenery, and for its history and legends. Its banks are covered with vineyards, many of its mountains and crags are crowned with ancient castles, and close to the river nestle picturesque towns and villages.

12. The scenery most admired is between Mayence, or Mainz, and Coblentz, a distance of about 50 miles, a part of which is seen in the picture.

13. The population of the German Empire is the same as that of the United States—40,000,000; and its area is the same as that of the four States, Michigan, Wisconsin, Illinois, and Indiana.

14. Berlin, the capital and largest city, has about the same population as Philadelphia and San Francisco combined—825,000.

HISTORICAL GEOGRAPHY.

1. **GERMANY** formed a part of the Empire of the West under Charlemagne, who was the first of a line of emperors extending through a thousand years ending A. D. 1800.

2. **The early inhabitants** were of a warlike nature; hence their name, German, which means war-man.

3. **The Germanic Confederation** of 1815 consisted of the several kingdoms, duchies, and free cities in Germany.

4. **By the war between Prussia and Austria** (in 1866), Austria was compelled to withdraw from the confederation. Prussia also annexed the Danish provinces of Schleswig, Holstein, and the German States Hanover, Nassau, Hesse Cassel, and the free city of Frankfort. The North German Confederation was then formed, consisting of all the States north of Frankfort.

5. **War between Prussia and France** was declared by Napoleon III (in 1870), who was defeated and taken prisoner. All the German States were then united to form the German Empire, with Prussia as the leading power.

6. **Poland** (meaning Flat Land), once a flourishing kingdom, now forms part of Prussia, Russia, and Austria.

AUSTRIA. What countries north of Austria? East? South? West? Southwest? What sea southwest? Which is the most western province? Northwestern? Northeastern? Southeastern? Which is the largest division of Austria?

Which is the principal river? Name its branches. Through what provinces does the Danube flow? What river on the northern border? On the southern? Where are the Carpathian Mountains? What river flows through Bohemia?

Where is Vienna (the capital and largest city)? Prague (the seat of Bohemian manufactures of glass, jewelry, etc.)? Pesth? Trieste (the great seaport of Austria)? Cracow (the ancient capital of Poland, and near extensive salt mines.

The area of Austria is about three times that of Kansas. (See page 98.) In population, Austria is nearly equal to the United States.

PHYSICAL AND DESCRIPTIVE GEOGRAPHY.

1. **AUSTRIA** is a mountainous country, with the great plain of Hungary in the center and south.

2. **Austria is rich** in agricultural and mineral products.

3. **The Hungarian provinces** comprise Hungary and Transylvania, and those south of Hungary.

4. **The German provinces** are between Hungary and the western boundary of Austria. Name them.

5. **The Polish province** is Galicia.

6. The precious metals are found in Hungary, Transylvania, and Bohemia; quicksilver and tin in the southwest; coal, iron, copper, and salt. in nearly all the provinces.

The City of Buda, or Ofen, is on the left of the picture, and the City of Pesth on the right.—View of the Danube, looking north.—Point to these cities on the map.

7. **Its exports** are of great variety; they include grain, cattle, wine, wool, salt, linen and leather goods, glass-ware, wooden-ware, gloves, instruments, and machines.

8. **The salt mines** in the north are the most celebrated in the world.

9. **The races** and languages are various.

Population of Austria, 36,000,000

HISTORICAL GEOGRAPHY.

1. **The word Austria,** which signifies *eastern state,* originally referred to what is now the provinces of Upper and Lower Austria, when they formed the eastern part of the dominions of Charlemagne.

2. **Austria** has, at various times, held and lost important possessions in Europe.

3. **It retains** the kingdoms of Bohemia and Hungary, and the province of Galicia which formed part of the Kingdom of Poland until it was seized by Austria, Prussia, and Russia.

4. **In the war** of 1866, Austria was defeated by Prussia and Italy, compelled to give up Venetia to Italy, and to withdraw from the Germanic Confederation.

SWITZERLAND.

What grand duchy north? What Austrian province east? What Italian states south? What country west? What lakes on the boundary of Switzerland?

What large river flows through Lake Constance? Through Lake Geneva? In what part of Switzerland do they both rise?

Name the largest lakes in Switzerland.

What river with its branches drains the greater part of Switzerland? What mountains in the south? In the northwest? Center? East?

Geneva, the largest city, is celebrated for watches, jewelry, and music-boxes. Where is it? Bern is the capital. Where is it? Where is Basel (*bah'zel*), or Basle (*bahl*)? Lausanne? Zurich (*zoo'rik*)? Fribourg? Neuchatel (*nush-ah-tel'*)? Lucerne?

Switzerland, like Holland, is about 1 m. long and ½ m. broad (100 x 200 miles).

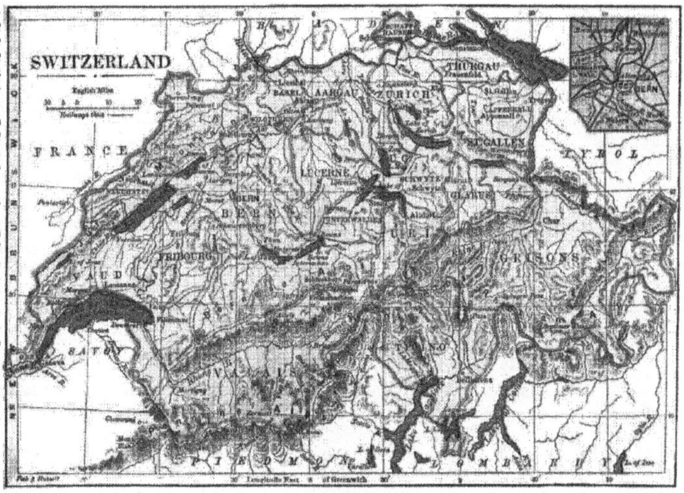

PHYSICAL AND DESCRIPTIVE GEOGRAPHY.

1. **Switzerland** is the most mountainous country in Europe. Its highest peaks are covered with perpetual snow.

2. **The scenery** of its mountains, glaciers (*gla'seers*), valleys, lakes, and waterfalls is grand and picturesque. (See illustrations on pages 6 and 8.)

3. **The highest peaks** are Mt. Rosa and Mt. Cervin, which are about 15,000 feet above the level of the sea. Mt. Blanc, the highest in Europe, is in France.

4. **The climate** is subject to great extremes of heat and cold.

5. **The Swiss** are hardy, industrious, and brave.

6. **They are employed** chiefly in dairy-farming and manufacturing. Cows, goats, and sheep constitute the wealth of the peasantry.

7. **The manufactures** consist chiefly of watches, jewelry, silk and cotton goods, and carved wood.

8. **The language** spoken in the central and northeastern cantons is German; in the western, French; and in the southern, Italian.

9. **Education** is general and compulsory; nearly every boy and girl can read and write.

10. **Switzerland** is a republic composed of cantons or states.

11. Population of Switzerland, 2,670,000—about the same as Illinois.

Fribourg or Freybourg, Switzerland.—Suspension Bridge over the Saane or Sarine River.

HISTORICAL GEOGRAPHY.

1. **At the beginning of the Christian era,** the inhabitants, the Helvetians, were subdued by the Romans, and in the 6th century by the Franks.

2. **The name** Switzerland is derived from that of the canton Schwytz, because its inhabitants distinguished themselves in securing the freedom of the country from Austria (14th century).

3. **The confederation** then comprised but three states, now in the central part of the Republic.

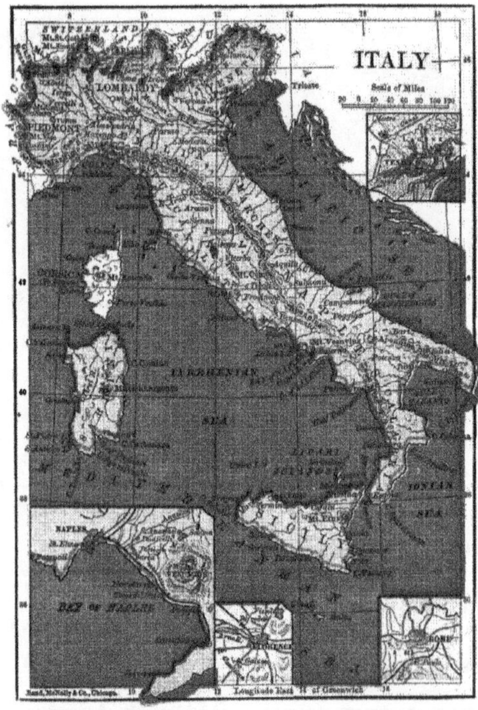

PHYSICAL AND DESCRIPTIVE GEOGRAPHY.

1. **ITALY** is celebrated for its mild winters, clear sky, and fine scenery, its volcanoes, ancient ruins, and works of art.

2. The three famous volcanoes of Italy are Vesuvius, Etna, and Strom'boli. Tell where each is situated.

3. **The north** is inclosed by several divisions of the Alps, whose high summits are covered with perpetual snow.

4. **The Apennines** begin at the Maritime Alps, near Genoa, and extend through the peninsula.

5. **On account of its shape** and the positions of its mountain chains, Italy has but one large river, the Po, which drains the rich and populous plains of Sardinia and Lombardy.

6. **The Po River** has numerous branches on its northern side, many of which drain lakes remarkable for the beauty of their scenery.

7. **The principal lakes** in the north are Garda, Como, and Maggiore (*mahd-jo'ra*).

8. **The western coast**, for a considerable distance north and south of Rome, is low and very unhealthy.

9. **The mildness of the winters** of Central and Southern Italy is chiefly due to the influence of the winds which blow over the waters of the Mediterranean Sea.

10. **The products** include grain, silk, wine, olive-oil, fruits, marble, and sulphur.

11. **The Italians** are a mixed race, from the Greeks, Gauls, Goths, Germans, and Arabs. They have a dark complexion, black hair and eyes. Many are refined and well educated. Schools, academies, universities, and libraries are numerous.

12. **San Marino**, a small republic, is northeast of Florence.

ITALY is a peninsula. The distance from Mt. Blanc, on the northwest, to the Strait of Messina, on the south, is 700 miles—about the same in length as the State of California.

What country north of Italy? Northeast? Northwest? What mountains on the northern border? Northeastern? Northwestern? What chain in the peninsula? *Ans. The Apennines.*

What sea east? West? Southeast? What gulf southeast? Northwest? What is the most southern cape?

Name the northern divisions,—the central and southern. What two large islands west of Italy? Which belongs to France? What British island south of Sicily? On what coast is Naples? Gen'oa?

Where are the Lipari Is. (*lip'a-re*)? Where is Rome, the most celebrated city in Europe? Milan? Genoa, the birthplace of Columbus? Venice? Naples, the largest city in Italy? Florence, celebrated for its valuable collections of sculptures and paintings? Leghorn, an important commercial city in Italy?

Observe that Mt. Blanc, Genoa, Rome, and Naples are nearly in a line with each other, and that the city of Naples is in the same latitude as New York.

The population of Naples is about half that of New York. Population of Italy, 27,000,000.

Rome.—The Tiber River, looking down the stream.—St. Peter's Cathedral and Castle of St. Angelo.

A view in the city of Venice, which has canals instead of streets, and gon'dolas or boats instead of carriages, etc. All the principal houses are built on the sides of the canals.

HISTORICAL GEOGRAPHY.

1. **ITALY** was settled by Greeks in the eighth century, B. C. For a long time its cities had separate governments, but all came under the dominion of Rome. (For Ancient Geography and maps, see pages 90 and 91.)

2. **Rome**, according to tradition, was founded by Romulus, about twenty-six centuries ago. It was a kingdom about 300 years, then a republic about 500 years, and became an empire in the year 35, B. C.

3. **The first Emperor after Cæsar**, was Augustus, who reigned at the beginning of the Christian era, when Rome was most celebrated for its wealth, power, splendor, and learning. Its dominion extended over nearly the whole of Europe, Western and Southwestern Asia, and Northern Africa.

4. **Constantine** was the first emperor who was converted from paganism to Christianity (300, A. D.).

5. **The States of Italy** have at times been under the dominion of France and Austria, but all are now joined together, forming the Kingdom of Italy.

6. **Pompeii** (*pom-pay'e*) and Herculaneum, ancient cities near Naples, were buried under the ashes sent out from Vesuvius during an eruption (in 79, A. D.).

A view in the city of Naples. The streets are narrow and paved with blocks of lava. The houses are high, gloomy, and crowded together, and the churches are remarkable for their size and their valuable works of art.

A Swedish home.—Reading and knitting, favorite employments during the long winter evenings.

SWEDEN, NORWAY, AND DENMARK.

Draw Sweden, 1 m. wide, 4¾ ms. long. (See p. 58.) Complete the map of Sweden and Norway. The peninsula of ***Jutland*** is ½ m. from north to south.

Into what do the rivers of Sweden flow? What lakes in Sweden? What water between Norway and Denmark? Sweden and Denmark? Bound Sweden,—Norway,—Denmark. Mention their capitals. Where is North Cape? The Naze? North Sea? The Loffo'den Is.? Gothland?

PHYSICAL AND DESCRIPTIVE GEOGRAPHY.

1. **The kingdoms of Sweden and Norway,** now under one monarch who resides in Stockholm, form the Scandinavian Peninsula; each has its own legislature.

2. **Their characteristics** are their indented coast lines with numerous islands, their forests of pines and firs, and their long winters; besides these, Norway is noted for mountains, glaciers, and snowy peaks, and Sweden for numerous rivers and lakes.

3. **Lumber, fish, and ice** are largely exported.

4. **Denmark comprises** the peninsula of Jutland and the islands east of it. Greenland and Iceland belong to Denmark.

Interior view of a fisherman's cottage in Lapland.

7. **At Hammerfest,** the most northern town in Europe, the sun in summer is seen until midnight.

8. **The established religion** in these three countries is Lutheran. Their languages are different, although derived from the same language, which is still spoken in Iceland.

9. **The Laplanders** live in the most northern parts of Europe; they are copper-colored and short, the men being less than five feet in height. They wear furs, live in huts which resemble bakers' ovens, and subsist on the milk and flesh of reindeer, which constitute their entire wealth.

10. Population of Sweden, 4,000,000; of Norway, 1,700,000; of Denmark, 1,800,000.

HISTORICAL GEOGRAPHY.

1. **The early history** of these three countries is obscure.

2. **The people were converted** from paganism to Christianity about the 10th century, and were sometimes allied and sometimes at war with each other.

3. **The most celebrated sovereign** of Sweden was Gustavus Adolphus, who reigned in the early part of the 17th century.

5. **The Swedes,** Norwegians, and Danes have a light complexion, and are strong, industrious, and hospitable. Nearly all can read and write, their children being compelled by law to attend school; they are employed chiefly in agriculture and the rearing of cattle and horses.

6. **Copenhagen,** on the island of Zealand, is the largest city. Stockholm is built on several islands.

Christiansund, a seaport in the southern part of Norway. Shipbuilding and an export trade in lumber, lobsters, dried and salted fish (chiefly herring and cod), are carried on.

4. **Norway** was joined to Sweden in 1814.

5. **Denmark** extended south to the Elbe until 1864, when the duchies of Sleswick, Holstein, and Lauenburg were taken by Prussia.

6. **The Danes** were long noted for their daring, especially on the sea; and, in the 11th century, they held England and a part of Scotland in subjection. (See page 62, paragraph 6.)

Russia and the Russians.

RUSSIA, TURKEY, AND GREECE.

What five countries of Europe border on Russia? What ocean and sea north? Three seas south? Sea and three gulfs west? What mountains and rivers on its boundary? What rivers in Russia flow north? Southeast? South? West? Between what gulf and lake is St. Petersburg? What peninsula in the north? South? Where is Moscow? Warsaw? Archangel? Odessa?

Bound Turkey. Mention its capital,—its mountains,—its rivers. What straits connect the Sea of Marmora with the Black Sea and Archipelago?

Bound Greece. What islands near it? What peninsula in the south?

PHYSICAL AND DESCRIPTIVE GEOGRAPHY.

1. **The characteristics of RUSSIA** are its vast plains, forests, and marshes; cold and barren in the north, but highly productive of grain in the center and south.

2. **The animals** comprise horses, cattle, sheep, goats, bears, and beavers, besides reindeer in the north, and camels in the south.

3. **The exports** include leather, hemp, flax, tallow, timber, and furs.

4. **The Russians** are generally uneducated. They are stout and strong, and have brown or sandy hair; their houses, which are built of wood, lack comfort and cleanliness. The Russians and Turks are civil and very grave in manner.

5. **The TURKISH or OTTOMAN EMPIRE** has possessions in Europe, Asia, and Africa.

6. **Turkey in Europe** consists of mountains, valleys, and plains. The forests are extensive.

7. **Among the leading exports** are grain, wool, cotton, goats' hair, and opium.

8. **The people** are of different races. The Mohammedans, who are the ruling class, constitute one-third of the population, and speak the Arabic language.

9. **The Turks** are proud, hospitable, and indolent, spending much of their time in smoking long pipes and sipping coffee. The women of the upper classes, when they appear in the streets, have their faces closely veiled.

10. **In Turkey and Greece** the schools are attended chiefly by boys.

Constantinople, which was founded by Constantine, the first Christian emperor of Rome, is the third city in size in Europe, having about the same population as New York. What cities in Europe are larger?

11. **GREECE** is the most southeastern country in Europe.

12. **Its characteristics** are its mountainous surface, beautiful scenery, mild dry climate, and ruins of ancient art.

13. **Its exports** comprise currants, lead, silk, figs, olive-oil, wine, bees-wax, and tropical fruits.

14. **The Greeks** are active and gay; they have dark eyes and hair and an olive-colored complexion. They are largely engaged in agriculture and pasturage. Their principal crop is currants.

15. **The government** is a constitutional monarchy (see page 13, paragraph 22.) The language resembles the ancient Greek.

16. **The people** of these three countries are mostly members of the Greek church (Christians).

Pop. of Russia in Europe, 70,000,000; Turkey, 18,000,000; Greece, 1,460,000.

HISTORICAL GEOGRAPHY.

1. **GREECE** was the first civilized nation in Europe. (See p. 91.) It was in the height of its power in the 4th and 5th centuries, B. C., when its famous battles were fought, and its celebrated poets, philosophers, orators, and sculptors appeared. Wealth, luxury, jealousies, decay, and subjugation followed in turn. It was taken by the Romans in the 2d century, B. C., and by the Turks in the 15th century, A. D. It became again independent in 1829, A. D.

2. **The most celebrated ruler of Russia** was Peter the Great, who founded St. Petersburgh nearly 200 years ago.

3. **Moscow** was captured by Napoleon (in 1812), but the city was set on fire, and the French compelled to retreat with great loss.

4. **The war** between Russia and Turkey (in 1854-'5) was caused by the invasion of Turkey by the Russians, who were defeated. The Turks were aided by France, England, and Sardinia.

Athens, the capital of Greece. On the height, called the Acropolis or citadel, are the ruins of the Parthenon and other heathen temples, which were built about 400 years B. C.

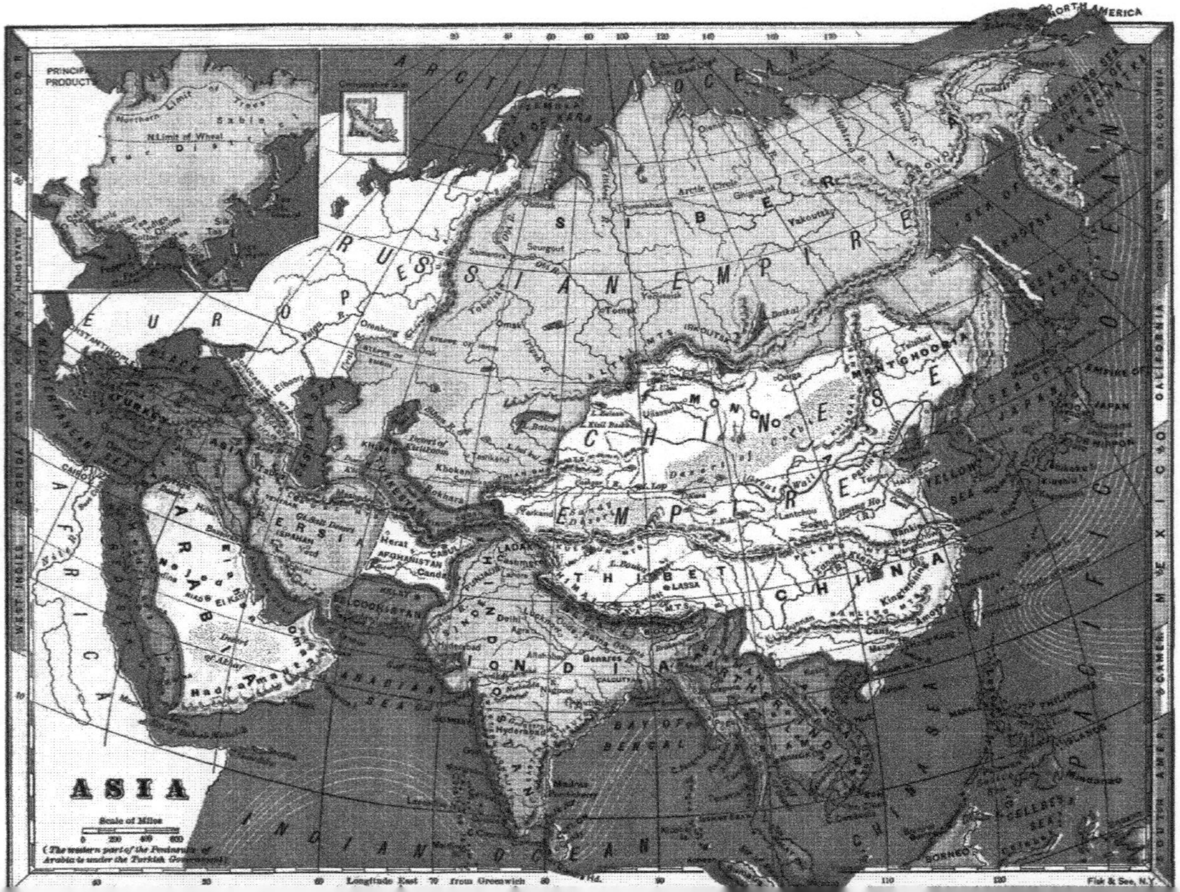
ASIA
Scale of Miles
(The western part of the Peninsula of Arabia is under the Turkish Government)
ARCTIC OCEAN
PACIFIC OCEAN
INDIAN OCEAN
RUSSIAN EMPIRE
SIBERIA
CHINESE EMPIRE
MONGOLIA
MANTCHOORIA
THIBET
CHINA
INDIA
PERSIA
ARABIA
AFRICA
EUROPE
TURKEY
AFGHANISTAN
BELOOCHISTAN
BAY OF BENGAL
ARABIAN SEA
YELLOW SEA
SEA OF JAPAN
NORTH AMERICA
PRINCIPAL PRODUCTS
Longitude East from Greenwich
Fisk & See, N.Y.

ASIA.

1. **ASIA** is the largest of the grand divisions.

2. **Its greatest extent** is from Behring Strait to the Strait of Bab-el-Mandeb.

3. **In area** it is equal to North and South America combined.

4. **About two-thirds of its area** is included in Siberia and the Chinese Empire.

5. **It has the highest mountains** on the Globe—the Himalayas. The highest peak is Mt. Everest, 29,000 ft.

GENERAL QUESTIONS.

By what oceans is Asia bounded? By what Grand Divisions?

What country forms the northern part of Asia? To what empire does Siberia belong? What large empire south of Siberia? What countries are included in the Chinese Empire? What countries are included in India?

What countries northwest of Hindoostan? Where is the nearest approach of the Russian Empire to Hindoostan? To North America?

What seas east of Asia? What two gulfs open into the China Sea?

Between what bodies of water is Hindoostan? Persia? Turkey? Farther India?

What peninsula in the eastern part of Siberia? What peninsula forms the most southern part of Asia? Between what seas is Kamtschatka? Corea? In what direction do these peninsulas point? What sea east of the Caspian Sea?

Through what countries does the Tropic of Cancer pass? What countries are partly in the Torrid Zone? Through what country does the Arctic Circle pass? What countries are wholly in the North Temperate Zone?

What mountains north of Hindoostan? What large rivers rise in the Himalayas? What mountains south of Siberia? What rivers rise in them? What mountains in Persia? Turkey? China? Thibet? Along the coasts of Hindoostan?

Toward what ocean do nearly all the rivers of Siberia flow? Name the largest rivers in Siberia. What three large rivers in Asia flow toward the Pacific Ocean? What large rivers flow through Hindoostan? Farther India? What rivers flow into the Aral Sea? What countries contain deserts?

REVIEW QUESTIONS.

Where are they? On or near what water?

Cities:—Pekin, Tokio (Yedo), Calcutta, Bombay, Madras, Hydrabad, Bangkok, Smyrna, Damascus, Mecca, Bokhara, Ispahan, Tabriz, Cabul, Kelat, Khiva.

To draw a map of Asia, see p. 96.

Where are they?

Islands:—Hondo, Luzon, Yeso, Borneo, Sumatra, Ceylon, Hainan, Formosa, Kurile Is., Kiushiu (*ke-oo'she-oo'*), Shikoku.

Mountains:—Himalaya, Altai, Stanovoy, Cau'casus.

Rivers:—Yang tse Kiang, Hoang Ho, Gan'ges, Mekong or Cambodia, Amoor', Lena, Yenisei, Obi, Tigris, Euphrates.

Questions on the Relief Map.—In what directions do the principal mountain chains extend? *They inclose table-lands of great elevation.*

What seas have no outlet to the ocean? What parts of the surface of Asia are low and level? What can you say of the surface of India? Of Arabia? Of Siberia? Of the Japan Is? Of Borneo and Sumatra? Of Persia? Of Turkey in Asia?

What parts of Siberia are mountainous? Hilly? Flat?

Has Arabia any large rivers?

1. **Northern Asia,** or Siberia, is remarkable for its great lowland plains, which slope gradually from south to north.

2. **Its great plains** contain extensive forests, steppes or prairies, marshes, fresh and salt lakes.

3. **In the south** are mountains and valleys.

4. **The winters** are long and intensely cold; the summers are short and hot.

5. **In the northern part** of Siberia the soil three feet below the surface is perpetually frozen.

6. **Yakutsk,** on the Lena River, the center of the fur-trade of Eastern Siberia, is the coldest city in the world, and Mecca, in Arabia, is the hottest.

A view in Northern Asia.—The Steppes or Prairies of Siberia.

7. **Siberia is rich** in metals, precious stones, and fur-bearing animals—sable, ermine, marten, beaver, bear, etc.

8. **The white inhabitants** are chiefly Russian settlers, criminals, and exiles from Russia.

9. **The natives** are idolators and very degraded; they have no settled habitations, subsisting chiefly by fishing and hunting. In winter they live in huts under ground.

10. **Siberia is a part** of the Russian Empire, whose capital is St. Petersburg, in Europe.

1. **India.** What rivers in Hindoostan? In Farther India? Where is Bengal? Sinde? Punjaub? Birmah? Siam? Anam? Nepaul? Bootan? Calcutta? Madras? Bombay? Cashmere? Mandelay? Bangkok? Hue? Gulf of Tonquin? G. of Siam? G. of Cambay? G. of Cutch?

2. **India is remarkable** for its hot, wet, and unhealthy climate, and the richness of its productions.

3. **Its surface** consists of mountains, plateaus, and lowlands.

4. **In the Himalayas** are the sources of the Ganges, Indus, and Brahmaputra.

5. **British India** includes Hindoostan and the east coast of the Bay of Bengal.

6. **The lowlands** are chiefly in the valleys of the Ganges and Indus Rivers.

7. **The rain-fall** is immense on the western coast and along the Brahmaputra.

8. **The Valley** of the Ganges is celebrated for its fertile soil and dense population.

9. **The trees** of India comprise the teak, the coconut, bamboo, banyan, tamarind, and palm.

10. Teak wood is valuable for shipbuilding, and the cocoanut is useful in many ways; of its leaves the natives thatch their houses; its fibres are made into matting, brooms, and baskets; its sap is used for drink, and its nut for food.

11. **The chief exports** are cotton, opium, dyes, drugs, spices, rice, silk, carpets, and shawls.

12. **The wild animals** are numerous; they include the elephant, rhinoceros, buffalo, lion, bear, tiger, panther, leopard, and monkey. Crocodiles and serpents also abound.

13. **The Hindoos** are of a dark complexion and have straight black hair; they belong to the Caucasian race.

14. **Their religion** is Brahminism, a form of idolatry.

15. **The population** is over four times that of the United States.

16. **Calcutta,** the capital, is the chief commercial city in Asia.

17. **The term India** often refers to British India alone. Cashmere, Nepaul, and Bootan, are independent.

18. **Farther India,** or Indo-China, is composed chiefly of the independent governments of Birmah, Siam, and Anam; its western part belongs to Great Britain. The Malay Peninsula is the most southern part of Asia.

A View in Southern Asia.—India.—The Himalayas, the highest mountains on the Globe.—Source of the Ganges River.

1. **SOUTHWESTERN ASIA** (comprising Arabia, Turkey, Persia, Turkestan', Afghanistan', and Beloochistan') is remarkable for its plateaus, deserts, and hot summers.

Questions on the Map of Asia.—Bound Turkey in Asia. What is that part called which is east of the Mediterranean Sea? East of the Red Sea? Describe its two large rivers.

Where is Smyrna? Jerusalem? Mt. Sinai? The island of Cyprus? Bound Arabia. Where is the Strait of Bab-el-Mandeb? Strait of Ormus? Mocha? Muscat? Sana (*sah-nah'*)?

Bound Persia. Where is Teheran (*teh-her-ahn'*), its capital? Mt. Ararat? Where are the Elburz Mountains? Has Persia any large rivers? Is the surface of Persia high or low? (See Relief Map of Asia.)

What and where is the capital of Afghanistan? Beloochistan?

2. **The rainless or desert region** of Asia extends from the Red Sea to the northeastern part of the Chinese Empire. Where is the Desert of Gobi, or Cobi?

3. **The tribes** and languages of these countries are numerous; their governments are despotic, and the prevailing religion is Mohammedanism. The languages most in use are the Arabian and Persian.

4. **Many of the inhabitants** are a wandering people, whose property consists of sheep, goats, horses, and camels.

5. **Inland trade** is largely carried on by means of camels in large companies, called caravans; but between Persia and Russia, it is by way of the Caspian Sea.

6. **The chief exports** are silk fabrics, shawls, carpets, fruits, wool, hides, etc. Besides these, Arabia produces coffee, grain, dates, and medicines, and in Persia are large deposits of salt.

7. **The chiefs** of nomadic (wandering) tribes are called sheiks; the sovereign of Persia is called the shah.

8. **The Afghans** are warlike and semi-barbarous.

9. **The Bedouins** are fierce, warlike, dishonest, and revengeful; they dwell in tents, and move from place to place to find pasture for their flocks and herds.

10. **The inhabitants** of Western Asia are chiefly of a brown complexion, yet they belong to the Caucasian race. The men wear long gowns or mantles. The women are made to work hard, and are but little respected.

11. **The largest cities** in Western Asia are Smyrna, Tabriz, and Damascus. Where are they?

12. **Turkestan** is chiefly under the dominion of Russia.

13. **Damascus** is the oldest city in the world; it was formerly celebrated for the manufacture of sword-blades.

14. **Mecca** is the birth-place of Mohammed, and is considered holy by his followers.

The creed of the Mohammedans or Moslems is called Is'lamism. It claims that Mohammed was God's prophet, and requires washings, fastings, almsgiving, sobriety, pilgrimage to Mecca, and praying five times every day with the face toward Mecca.

15. **Mocha** is celebrated for coffee. Where is Mocha? Mecca?

16. **Arabia** comprises desert tracts, with rich oases, some of which yield luxuriant vegetation and support crowded populations.

17. **The population** of the Wahabee empire, in Central Arabia, is more than a million. Its capital is Ri'ad.

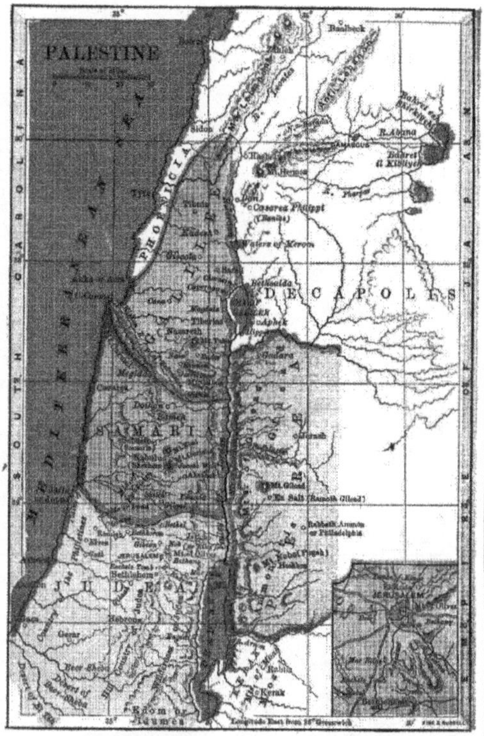

PALESTINE. What were its four divisions? What sea west? What is the largest river? Through what sea does the Jordan flow? Into what sea does it empty?

What large city in the northern part of Judea? What city south of Jerusalem? What mount and village east? (See corner of map.) Where is the site of Jericho? What city on the northwest coast?

Name the principal places in Samaria.

In what part of Galilee is Nazareth? Mt. Tabor? Mt. Gilboa? Mt. Carmel? Tiberias? The site of Capernaum? Kadesh?

18. **Palestine,** the southern part of Syria, is under the dominion of Turkey. It is but little larger than Vermont.

19. **The surface** is level in the west, and high in the center. The Dead Sea and the lower valley of the Jordan are more than 1300 feet below the level of the Mediterranean Sea, and the climate of that region is intensely hot.

20. **The plains** of Sharon and Esdrae'lon are very fertile, producing grain, grass, and tropical fruits; but the hills of Judea and eastward to the Dead Sea are mostly barren and rocky. The government is inefficient, crops and other property are insecure, and agriculture is in a backward state. In some places, shepherds and travelers must be armed for protection against the attacks of Bedouins.

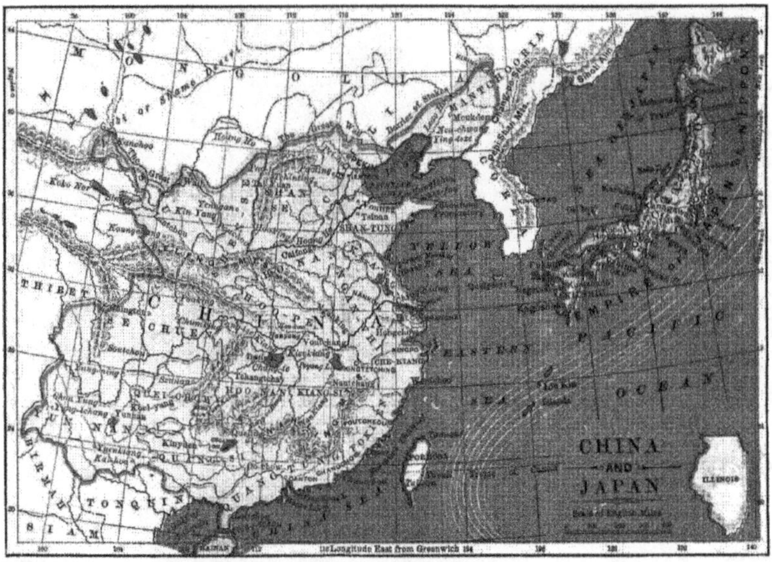

CHINA AND JAPAN:—What are the boundaries of China? What mountains in China? What seas east? What two large rivers flow through China? Into what do they empty?

What islands east of the Sea of Japan? East of China? South and southeast of China? What desert in the northern part of the Chinese Empire? What peninsula in the east? Where is Pekin? Canton? Nankin? King-tetching? Changchow? Tientsin? Soochow? Amoy? Shanghai?

Mention the Japan Islands. Which is the largest? Where is To'kio or Tokei?

What strait between Hondo and Yezo (or Jesso)? What channel south? In what part of Japan is Yokohama, its chief seaport? Nagasaki? Kioto? Hakodate? O'zaka?

What cities in the United States are in the same latitude as the northern part of Japan? (See margins.)

What parts of Asia and what cities in North America are between the parallels of 40° and 50° north latitude? (*See Maps of Asia and North America.*)

What parts of America, Europe, and Asia are between the parallels of 50° and 60° north latitude?

The parallel of 30°, which passes over New Orleans, passes over what parts of Asia?

1. **THE CHINESE EMPIRE** contains about one-third the population of the Globe.

2. **In its northern part** is a vast rainless region, the desert of Gobi or Cobi; but in Mantchooria and Chinese Turkestan there are large fertile tracts.

3. **China proper** forms one-third of the Empire. Its surface is high on the west and north, with rich and highly-cultivated plains in the center.

4. **The agricultural products** are tea, rice, cotton, sugar, grain, and fruits. In the south, olives, oranges, and pineapples are raised, and the mulberry tree is cultivated for the support of the silkworm.

5. **Its manufactures** comprise silk goods and porcelain ware.

6. **The mineral products** include gold, silver, copper, lead, iron, and coal.

7. **The climate** is colder in winter and warmer in summer than in corresponding latitudes in California and Western Europe. Why? (See p. 11.) In summer the heat is intense, with hurricanes, typhoons, and thunderstorms.

8. **The Chinese**, Japanese, and Tibetans belong to the Mongolian race, and are very singular in their appearance. They are small in stature. Their color is dark yellow, their eyes black, small, and obliquely set, and their dress, habits, and modes of agriculture and manufacture are similar to what they were thousands of years ago. They are industrious, peaceful, and have great respect for the aged, and veneration for the dead. The Japanese have adopted the European, or American, style of dress. The heads of Chinamen are shaven, except a long lock called a queue (*qu*), and the upper classes usually appear with umbrella and fan.

9. **The religions** of China are those of Buddha, Confucius, and Tao. Buddhism, which is the most prevalent, is now a kind of paganism.

10. **The people are gradually changing** their customs, owing to their intercourse with Americans and Europeans.

11. **The Chinese are addicted** to the use of opium, which is largely imported from Hindoostan and Asia Minor.

12. **The great wall** on the north was built for protection from their enemies, the Tartars (3d century, B. C.).

13. **Pekin**, the capital, is, next to London, the largest city in the world. It has twice as many inhabitants as New York.

14. **Canton,** with a population equal to that of New York (1,000,000), is the chief commercial city in the empire. It exports tea, silk, precious metals, sugar, and porcelain.

15. **TIBET,** or **THIBET,** is very high and dry. Vegetation and fuel are scarce.

16. **It manufactures** considerable cloth for China.

17. **THE EMPIRE OF JAPAN** or **NIPPON** is composed of islands, the largest of which are Hondo (*hone'do*), Yezo or Jesso, Kiushiu (*ke-oo'she-oo'*), and Shikoku.

18. **Its area** is about the same as that of California, which lies due east, and its population (33,000,000) is nearly equal to that of the United States. Its capital is Tokio (*to'ke-o*), formerly called Yedo.

19. **The climate** is tropical in the south, temperate and cold in the north.

20. **Japan contains** many mountains and active volcanoes.

21. **Agriculture** receives great attention. Most of the hills are cultivated to their summits. The products comprise grain, tea, cotton, sugar-cane, and tobacco, besides camphor and varnish.

22. **The principal food** of the inhabitants of Farther India, Japan, and the Philippine Islands consists of rice and fish.

23. **The mineral products** are gold, silver, copper, and coal.

24. **The islands are remarkable** for the frequency of heavy rains, fearful hurricanes, earthquakes, and volcanic eruptions.

25. **The Japanese** and Chinese belong to the Mongolian race.

26. **The Japanese manufactures** which are celebrated include sword-blades, watches, silk goods, camphor, varnish, porcelain, lacquer and japanned ware; in the latter, excelling every other nation in the world.

27. **The first treaty** between Japan and the United States was concluded by Com. Perry (1854).

28. **The earliest authentic accounts** of China and Japan published in Europe were given by Marco Polo, a celebrated traveler of the 13th century.

29. **Education** in China and Japan is encouraged and well rewarded.

30. **The Chinese language** is the oldest spoken language in the world.

31. **The Philippine Islands** (*fil'ip-pin*) belong to Spain. They export sugar, tobacco, hemp, and rope.

32. **The people** are chiefly Malays, many of whom are intelligent mechanics.

33. **Manilla** is the capital and chief seaport of the Philippines.

TOPICAL GEOGRAPHY.

EUROPE.

Where is Europe situated? What is its comparative extent? Where are its great plains? Its high mountains? Mention its principal rivers. What can you say of the Rhine?

What countries have a cold climate? A warm climate? A moist climate?

Ask the following questions about each of the countries in Europe:

ENGLAND,
SCOTLAND,
IRELAND,
FRANCE,
SPAIN,
PORTUGAL,
BELGIUM,
HOLLAND,
PRUSSIA,
GERMAN EMPIRE,
AUSTRIA,
SWITZERLAND,
ITALY,
SWEDEN,
NORWAY,
RUSSIA,
TURKEY,
GREECE.

What is its latitude or position on the Globe?

What can you say of its size?

Surface?

Climate?

Products and exports?

Manufactures?

Inhabitants?

Occupations of the inhabitants?

History?

ASIA.

What can you say of its size? Position on the Globe? Where are its great mountains? Plains? Deserts? What is the climate of Northern Asia? Southern Asia?

What can you say of the Inhabitants of Siberia? What are exported from Siberia?

Name the coldest city in the world. Where is it? Which is the hottest city in the world? Where is it?

What can you say of the climate of India? What are its principal products? Trees? Manufactures? Wild animals? Its inhabitants?

In what occupation are many of the inhabitants of Western Asia engaged? To what race do the people of Southwestern Asia belong? What is their complexion? Are their languages few or many? What is the style of their dress? What are the leading exports from the southwestern countries of Asia?

What are the people of Afghanistan called? What can you say of the Afghans? Of the Bedouins? Under what government is Palestine or the Holy Land? What are the largest and most celebrated cities in Western Asia?

What can you say of Arabia? Of the Valley of the Jordan?

What three countries in Asia are under European governments?

Mention the principal agricultural products of Eastern and Southeastern Asia,—manufactures. Which of the Asiatics belong to the Mongolian race? What can you say of the size of those people? Their color? Dress? Food? Disposition? Religions?

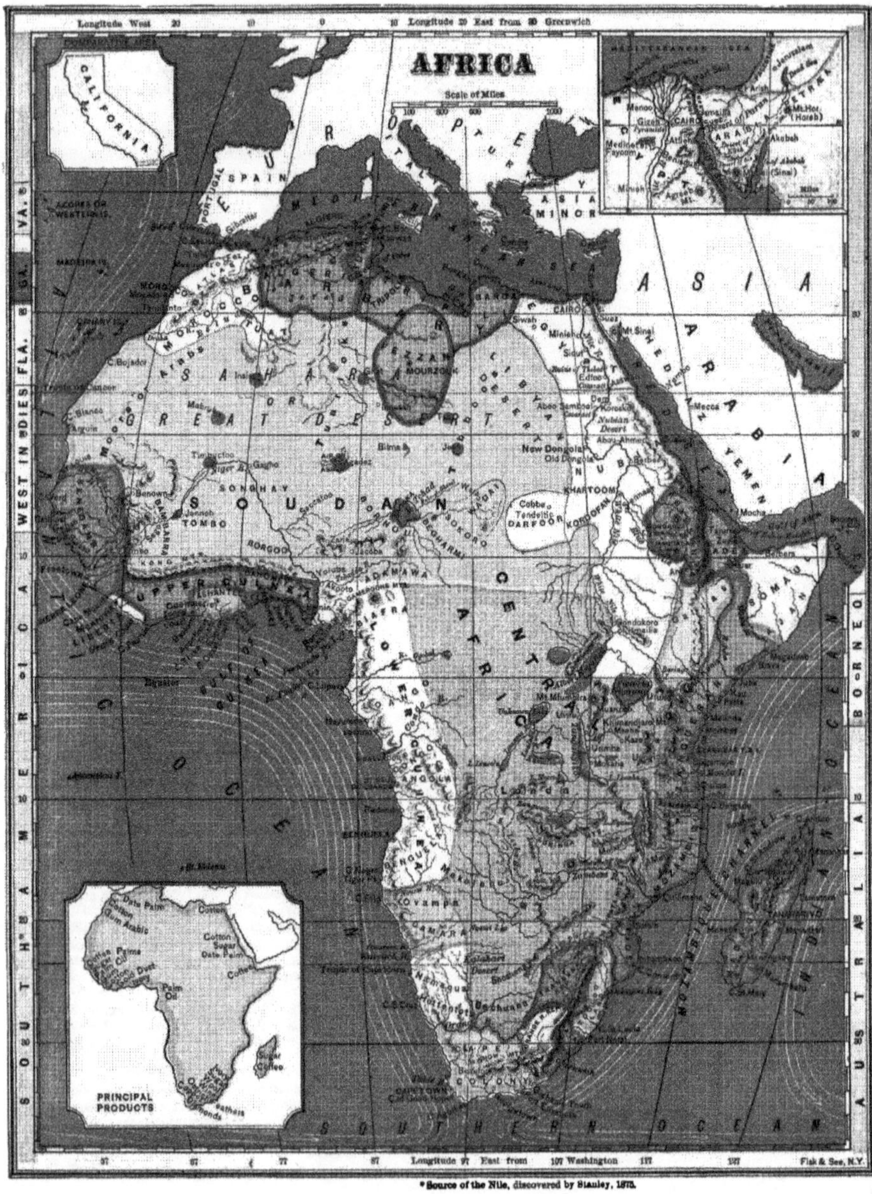

* Source of the Nile, discovered by Stanley, 1875.

AFRICA.

1. **AFRICA** is the hottest and most central division of the Earth; the Equator passes almost through its center.

2. **It is remarkable** for its high surface, extensive deserts, and hot climate.

3. **Its coasts** are low and unhealthy, but the interior consists of high table-lands. Africa is deficient in gulfs, bays, and other indentations of the coast.

Compare the coasts of Europe and the United States with those of Africa.

4. **Its northern part** is directly east of the Southern States.

5. **Its principal mountain chains** are near the coast.

REVIEW QUESTIONS.

Where are they?

Mountains:—Atlas, Kong, Cameroons, Mt. Kenia, Mt. Kilimandjaro.

Rivers:—Nile, Niger, Zambeze, Congo, Orange, Senegal, Blue, Coanza.

Capes:—Bon, Agulhas. Good Hope, Guardafui (*fwee'*), Verd, Blanco, Frio (*fre'o*), Palmas.

Islands:—Madagascar, Com'oro Is., Socotra, St. Helena, Azore Is., Madeira Is., Canary Is., Zanzibar.

Observe that the mountain ranges lie in the same general direction as the coasts, which are near them; that the highest mountains and lakes are on the eastern side, and near the Equator; that the sources of the Nile, Zambezi, and Congo Rivers are on the great table-lands of the interior, which are very high and abundantly supplied with rain; and that north of the central part of Nubia the Nile flows through a long, narrow valley, without receiving a single tributary.

GENERAL QUESTIONS.

By what is Africa bounded on the north? East? South? West? What cape at its northern extremity? Southern? Eastern? Western?

What large island southeast? What channel between it and the main land? What strait at the entrance to the Mediterranean Sea? At the entrance to the Red Sea? What gulf and island near Cape Guardafui (*gwar-dah-fwee'*)? What group of islands west of Cape Verd? *Cape Verd Islands.* What groups northwest of Africa?

What two high mountains south of the Equator? What range in Northern Africa? What other ranges in Africa?

What lakes near the Equator? What great river drains them? What large river in Western Africa?

What desert in the north? What countries border on the Mediterranean Sea? On the Red Sea? On the Gulf of Guinea? What countries between the Equator and the Tropic of Capricorn? What countries south of that tropic?

What city in the United States is in the same latitude as Cairo, the capital of Egypt? (See page 28.) What city in Africa is in the same latitude as Charleston?

Over what part of Africa does the Equator pass? The Tropic of Cancer? The Tropic of Capricorn? In what zone is the greater part of Africa? What countries are in the North Temperate Zone? The South Temperate Zone?

Bound Egypt. On what river is its capital? What city near the western mouth of the Nile? What town at the head of the Red Sea? At the junction of the Blue and the White Nile? In Fezzan? What is the capital of Morocco (or Marocco)? Algeria? Tunis? Ashantee? Dah'omey? Cape Colony?

What town and cape on the coast of Liberia? What town on the coast of Sierra Leone (*se-er'rah le-o'na*)?

What canal connects the Mediterranean and Red Seas? *Ans. The Suez Canal.*

By what two routes can you sail from Spain to the Indian Ocean?

What village on the eastern side of Lake Tanganyika?

For Map Drawing see p. 97.

RELIEF MAP OF AFRICA.

AFRICA.

1. **AFRICA** is chiefly a great table-land; its rainless or desert region is Sahara, in the north. South of the center, rain is abundant, supplying the Nile, Zambezi, Congo, and other rivers, besides extensive lakes.

2. **The principal lakes** in Central Africa are Victoria, Albert, Tanganyika, and Tchad.

3. **The productions** of Africa comprise grain, cotton, sugar, coffee, tobacco, indigo, ivory, ebony, ostrich feathers, palm oil, and tropical fruits.

A view in Northern Africa.—Tangier, a town in Morocco, near the entrance to the Strait of Gibraltar. The tops of the houses, which are flat, are the popular resort in the cool of the evening.

4. **The wild animals** are very numerous; they comprise the lion, elephant, rhinoceros, hippopotamus, leopard, giraffe, zebra, and monkey. The gorilla is found in the Equatorial regions, and the crocodile in the rivers.

5. **The Barbary States** are in the north. They comprise Morocco, Algeria, Tunis, and Tripoli. Barca and Fezzan belong to Tripoli. These countries are inhabited by Moors, Berbers, Arabs, and Turks—all Mohammedans and Caucasians, with straight black hair and of a dark complexion. (See page 66, Historical Geography, paragraph 2.)

6. **Cattle, horses, and goats** are numerous.

7. **Rains** are frequent in winter, but seldom seen in summer.

8. **The principal products** of Barbary are leather, wool, grain, and olive oil, with dates, olives, and other fruits.

9. **Morocco** is an empire; Algeria belongs to France; Tunis (ancient Carthage) and Tripoli are each governed by a Bey, who is subject to the Sultan of Turkey.

10. **Egypt** is in the same latitude as Florida. It has hot summers, mild winters, no snow, and but little rain.

11. **The Valley of the Nile** is celebrated for its fertility, due to the annual rise of the river Nile.

12. **In Egypt**, the river is inclosed between high banks, through which the water is conducted in narrow channels, and allowed to cover the land, leaving a rich sediment.

13. **The rise of the water,** caused by the abundant rains on the highlands of Eastern and Central Africa, begins at Cairo in June, and continues until September. Without the Nile, Egypt would be a desert.

14. **Considerable trade** between Europe and India is carried on through Egypt, by way of the canal and railroad which extend from the Mediterranean to the Red Sea.

15. **The inhabitants** of Egypt comprise Fellahs, Copts, Arabs, and Turks. The Fellahs are the peasants and laborers, and the Turks are the ruling class. All are Mohammedans, except the Copts, who are Christians. Besides these, many Europeans, Jews, and Syrians live in Egypt.

16. **The language** of the Egyptians is Arabic; their complexion is a brownish yellow. They belong to the Caucasian race.

17. **Cairo** is the capital and largest city. Alexandria, founded by Alexander the Great, is the principal seaport.

18. **Egypt is celebrated** for its magnificent pyramids, temples, obelisks, statues, and tombs built more than 4000 years ago.

Cheops (*ke'ops*), the great pyramid, is over 450 feet high, and its base covers an area of twelve acres.

19. **Nubia** is a desert, except the valley of the Nile and its southern or rainy section, where vegetation is abundant. It is under the dominion of Egypt, and its inhabitants are Arabs and blacks, all Mohammedans.

20. **Abyssinia** is noted for its high mountains and plateaus, hot valleys, and heavy rains in summer.

21. **The forests and pastures** are extensive, tropical productions abundant, and wild animals numerous.

22. **Abyssinia is divided** into states, which are despotically governed by chiefs.

A view in Abyssinia.—A chief and his warriors; servant-maids bringing fruit.

23. **The Abyssinians** profess Christianity, but are very superstitious.

24. **Sahara** is the largest desert in the world. It is nearly as large as the United States. It is a table-land, consisting of vast sandy flats and dry barren rocks.

25. **The oases** are fertile places, where springs of water, trees, and grass are found.

26. **Soudan**, or Nigritia, comprises a number of independent states. Its principal place is Timbuctoo (Where is it?), important for its trade between Guinea, Senegambia, and the Barbary States.

27. **The climate** is intensely hot, and in the rainy season, large tracts of land in Central Africa are inundated; it is, therefore, very unhealthy in summer.

28. **The Negroes** inhabit Soudan, Central and Southern Africa. They are chiefly pagans, and many are barbarous; but in Soudan the Negroes are largely engaged in agriculture.

29. **Liberia** is a republic of American Negroes.

30. **Sierra Leone** is a British colony.

31. **The British, French, and Portuguese** have settlements along the western coast.

32. **Cape Colony**, Natal, and Kaffraria are British colonies. Where are they?

33. **The white inhabitants** of Southern Africa are chiefly British and the descendants of early Dutch settlers. The latter rule in the Orange River Free State and the Transvaal Republic. Where are these countries?

34. **The Kaffirs** are a pastoral people, but brave and warlike. The tribes are governed by chiefs.

Moors of Northern Africa.—Moorish Architecture.

35. **The climate** of Southern Africa is delightful. Cape Good Hope and Cape Hatteras (Where are they?) are the same distance from the Equator, one in north, and the other in south, latitude: and when it is midwinter at one, it is midsummer at the other.

36. **Important discoveries** of diamonds have been made in Southern Africa. Ivory and gold-dust are exported from Southern and Eastern Africa.

Interior of a Negro Village.

37. **Sofala and Mozambique** belong to the Portuguese, and Zanguebar to the Arabs.

38. **The countries** in Eastern Africa, between the Equator and Abyssinia, are inhabited by the Gallas and other savage tribes.

39. **Among the celebrated explorers** of Africa are Barth, Du Chaillu, Grant, Speke, Baker, and Livingstone.

40. **Madagascar** is a monarchy of greater area than France. The inhabitants are chiefly employed in rearing cattle.

41. **The ancient** Egyptians and Carthagenians were powerful nations.

42. **Egypt** was civilized when Europe was in a state of barbarism.

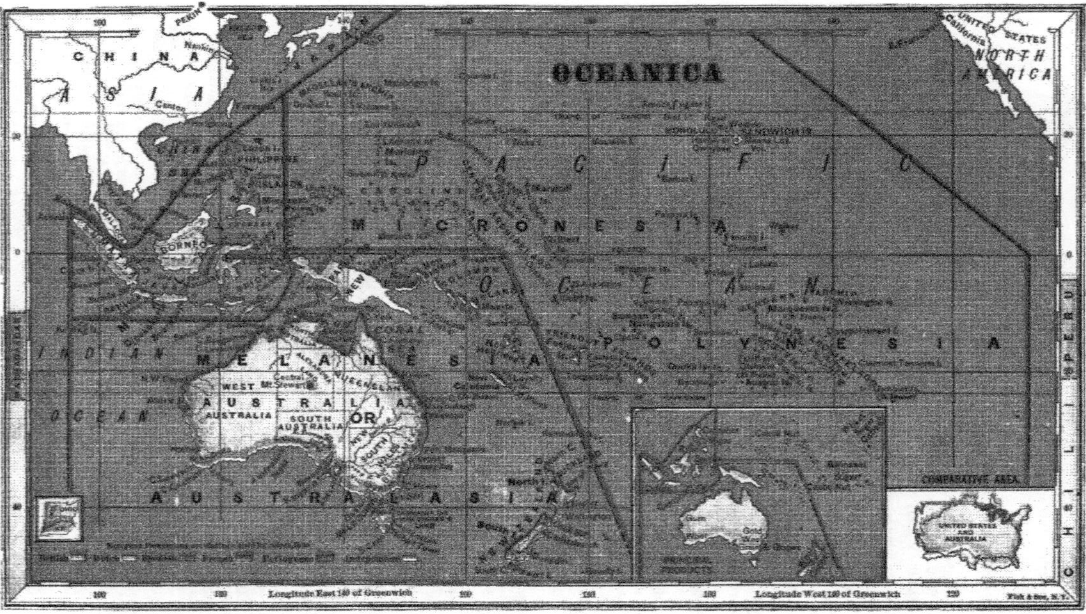

EXERCISES ON THE MAP.

What are the divisions of Oceanica?

Which contains the largest islands?

Micronesia signifies *small islands*; Polynesia, *many islands*.

Mention the largest island. What division is between Australia and Asia? Name the largest island in Malaysia.

Through which does the Equator pass? What towns in Sumatra?

What is the capital of Java? What island east of Borneo?

What group of islands in the northern part of Malaysia? In the eastern?

Name the divisions of Australia. What cities in the southeast?

What capes on the northeast? Southeast? Northwest? Southwest? Where is the gold region of Australia?

What separates Tasmania from Australia?

REVIEW QUESTIONS.

Where are they? By what waters are they surrounded? In what part of what Division are they?

Islands:—Java,	Mendana Arch.,	Ladrone Is.,
Sandwich Is.,	New Zealand,	Caroline Is.,
New Guinea,	Friendly Is.,	New Ireland,
Borneo,	Magellan's Arch.,	Feejee Is.,
Australia (*Aws-tray'le-a*),	Society Is.,	Spice Is.,
	Tasmania,	Luzon,
Philippine Is. (*fil'ip-pin*),	Sumatra (*su-mah'trah*),	Celebes (*sel'e-bes*),
Hawaii, (*hah-wi'e*),	Central Arch. (*ark-e-pel'a-go*),	Marquesas (*mar-ka'sahs*).

Where are they?

Cities and Towns:—Melbourne, Bencoolen, Adelaide, Acheen, Batavia, Auckland, Borneo, Sydney.

Straits:—Malacca, Tor'res, Sunda, Macassar, Cook's, Bass.

Seas, Gulfs, and Bays:—Coral Sea, Botany Bay, Gulf of Carpentaria, China Sea, Spencer's Gulf, Gulf of Cambridge, Java Sea, Celebes Sea.

In what direction is Australia from San Francisco? What islands on the route from San Francisco to China and Japan? Why not return by the same route? (*See Oceanic Currents, on page 11.*) What is the capital of the Sandwich Is.? *Honolulu.*

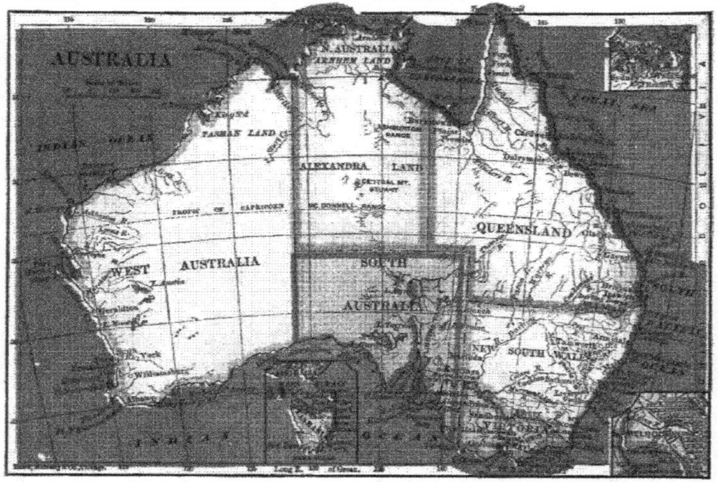

6. **The climate** and vegetable products of the northern part are tropical, while those of the southern part belong to the temperate zone.

7. **Australia** being in the southern hemisphere, its seasons are the reverse of those in the northern hemisphere.

8. **Australia, New Zealand, and Tasmania** belong to Great Britain.

9. **The inhabitants** are chiefly British, and their principal occupations are mining, agriculture, and grazing. The original inhabitants of Australia are short and stout, with small heads, flat noses, thick protruding lips, long coarse hair, and of a black or dark brown complexion.

10. **The original inhabitants of New Zealand,** called Maories, were formerly fierce cannibals; but they are now mostly civilized; they are active and well formed, with prominent features, black flowing hair, and of a copper complexion.

11. **The characteristics** of New Zealand are its mountains, evergreen forests, rich valleys, and heavy rains in their winter—July. (See page 10, paragraph 29.)

12. **Pap'ua,** or New Guinea, contains high mountains and extensive forests.

13. **Among its products** are camphor, sago, cocoanuts, nutmegs, and other spices.

14. **It is inhabited** by Negroes. Its western half is claimed by the Dutch.

To Draw Australia.—Mark Melbourne in the southeast; 3½ ms. north by west mark Cape York, and the same distance west by north, Cape Leeuwin. Complete the coast-line of Australia, and mark its divisions, with their chief towns, capes, gulfs, and rivers.

Is Australia north or south of the Equator? What tropic passes through it? In what zones is it? Mention its principal capes,—rivers,—its divisions or colonies. What is the capital of the colony of Victoria? Of New South Wales?

In what direction from Australia is New Zealand? California? What two large islands included in New Zealand? What strait between them? Where is Tasmania? What strait between Australia and Tasmania?

What large island north of Australia? What strait? What gulf?

What three large islands are crossed by the Equator? Where is Sunda Strait? Macassar Strait? Celebes Sea? Java Sea? What group of islands southwest of California, about one-third the distance from San Francisco to China or Japan? Which is the largest of the Sandwich Islands? What islands east of the China Sea? What island east of Australia belongs to France? What islands are east of New Guinea? Name the principal groups of islands north of the Equator,—south of the Equator. In what direction from San Francisco is Yedo? Pekin? Melbourne?

PHYSICAL AND DESCRIPTIVE GEOGRAPHY.

1. **OCEANIA** or **OCEANICA** comprises nearly all the islands of the Pacific Ocean.

2. **Australia** is about as large as the United States.

3. **Its mountains** are near the coast, and its interior is a plateau.

4. **In the southeast** are its highest mountains, largest rivers, and the most of its population and wealth.

5. **Its wealth** is chiefly in gold, copper, and wool.

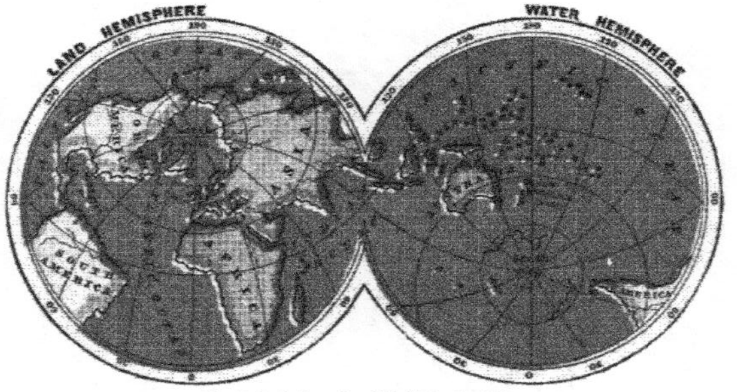

New Zealand, the center of the Water Hemisphere.

17. **The islands of Malaysia** are noted for their hot, moist climate, luxuriant vegetation, numerous volcanoes, and frequent earthquakes.

18. **The most important islands** are Borneo, Sumatra, Java, Celebes, the Philippine, and Spice Islands.

19. **The products** include ebony, gutta-percha, cloves, nutmegs, pepper, ginger, cinnamon, rice, cotton, tobacco, coffee, sugar, fruits, bamboos, and rattans.

A view on the coast of Borneo.

20. **From the bamboo,** the natives make houses, beds, bridges, baskets, etc.

21. **The animals** of Malaysia are the elephant, rhinoceros, tiger, panther, and monkeys.

22. **Orang-outangs** are found in Borneo and Sumatra.

23. **Birds** of Paradise, parrots, and other beautiful birds are very numerous.

24. **The inhabitants** are Malays and Papuan Negroes, besides Dutch, English, and Chinese settlers.

25. **Many of the Malays** are bold and piratical.

26. **Java** is remarkable for active volcanoes. Its area is nearly equal to that of England.

27. **The Sandwich Islands** are in the route of vessels trading between the United States and China, Japan, and Australia.

28. **They are remarkable** for the mildness of their climate, for earthquakes, and the volcano, Mauna Loa, on Hawaii (*hah-wi'e*), the largest island of the group.

29. **The inhabitants,** chiefly of the Malay race, are rapidly advancing in civilization.

30. **The government** of the Sandwich Islands is a kingdom, the capital of which is Honolulu.

GENERAL EXERCISES IN PHYSICAL GEOGRAPHY.

(See Charts on pages 89 and 11.)

On which side of the Equator is the most land? The most water?

What grand divisions and what islands are crossed by the Equator?

What part of South America is crossed by the Equator? What part of Africa?

What countries are crossed by the Tropic of Cancer? By the Arctic Circle?

Observe that certain coast-lines are parallel, or nearly parallel, with each other.

Extending from northeast to southwest are the eastern coasts of Asia, Africa, South America, the United States, and Greenland, the western coast of Europe, and the northwestern coast of Africa.

Observe that a line drawn northeasterly from the Strait of Magellan to North Cape of Europe would almost coincide with the east coast of South America, the northwest coast of Africa, and the west coast of Europe.

Where are the narrowest parts of the Atlantic Ocean? Where is the narrowest part of the Western Continent?

Name the Oceanic currents. Which is in the Torrid or hot Zone?

In what direction does the Equatorial current in the Atlantic flow?

What warm current or stream proceeding from the Equatorial current begins at the Gulf of Mexico and flows northeastwardly?

What coasts of Europe are washed by the water of the Gulf Stream? How far north does the northeast branch extend?

What effect has the Gulf Stream upon the climates of Western Europe?

What similar current east of Asia?

How far north is the vine cultivated? Wheat?

Observe that between the North American coast and the Gulf Stream is a cold current from the north.

Where is the whale found? The seal? Where is cotton cultivated?

Famous for gold mines are the United States and Australia; for tin, lead, iron, and coal, is Great Britain.

Mention some of the mineral products of North America,—South America,—Asia,—Australia. *The most famous silver mines are in Mexico.*

Mention some of the vegetable products of South America,—of Asia,—of Africa,—of Arabia,—of India. Where is wool largely produced? Rice?

Besides Louisiana, what parts of the earth produce sugar? Where does coffee grow? Tea? Where is silk obtained? Wine? Cinnamon? Pepper? Palm Oil? Opium? India-rubber? Peruvian Bark? Indigo? Ivory? Guano? Fur? Where are diamonds obtained? Hides and Tallow?

In what direction would you sail from the United States to the British Isles? Would you then sail with or against the Gulf Stream?

Which is the most numerous of the five races? *Ans. The Mongolian.*

Physical and Commercial Chart of the World.

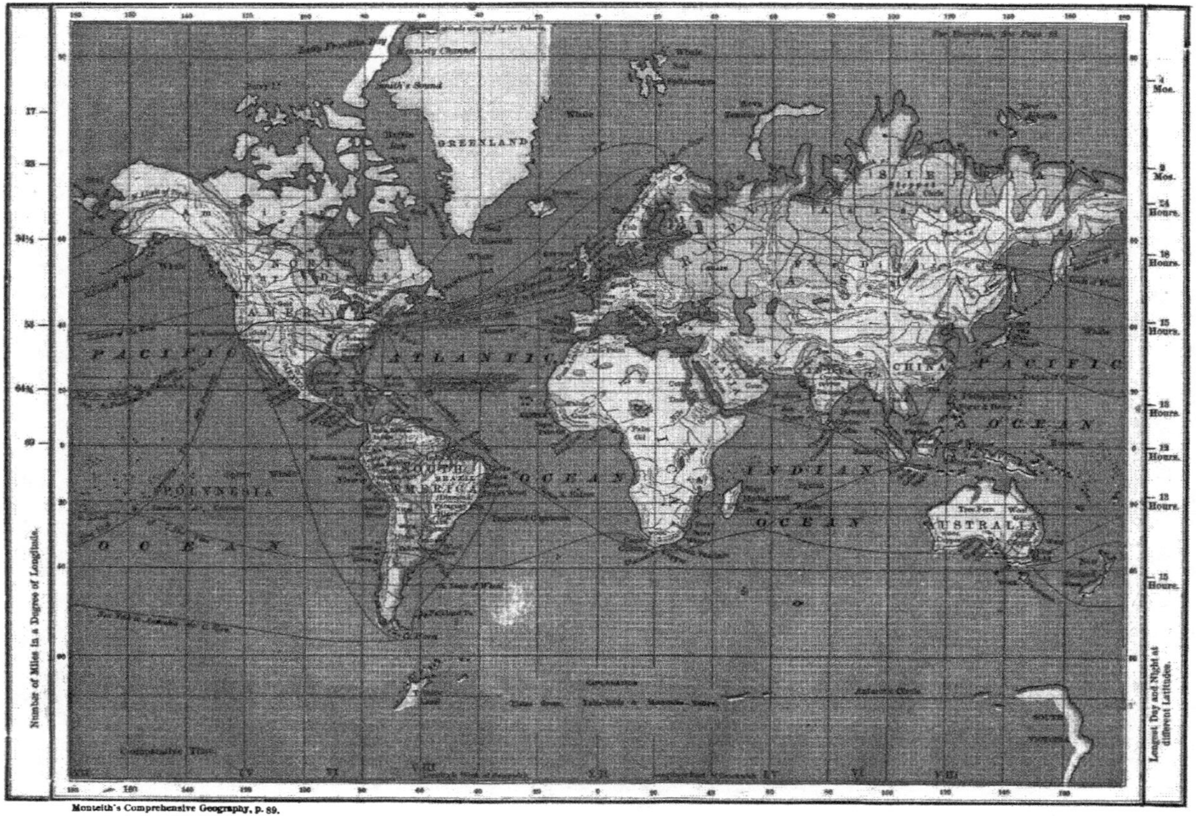

Monteith's Comprehensive Geography, p. 89.

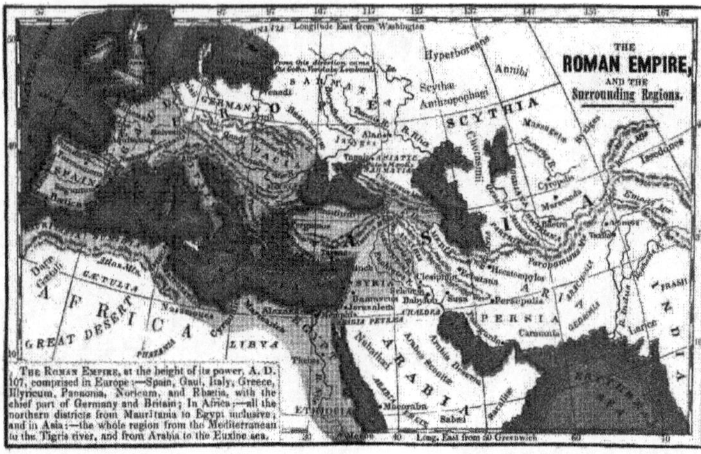

The Roman Empire, at the height of its power, A. D. 107, comprised in Europe:—Spain, Gaul, Italy, Greece, Illyricum, Pannonia, Noricum, and Rhaetia, with the chief part of Germany and Britain; In Africa:—all the northern districts from Mauritania to Egypt inclusive; and in Asia:—the whole region from the Mediterranean to the Tigris river, and from Arabia to the Euxine sea.

1. **The first settled** parts of the earth were in southwestern Asia and northeastern Africa.

2. **The Garden of Eden,** it is supposed, was situated near the Tigris River.

3. **The Deluge,** mentioned in the Old Testament, occurred more than 4,000 years ago; the ark in which Noah and his family were saved rested on Mount Ararat, in Armenia. (See page 13, Historical Geography.)

4. **Babylon,** founded about 150 years after the Deluge, was celebrated for its magnificence.

5. **Nebuchadnezzar** was king of Babylon in the year 600, B. C. He conquered Palestine and Egypt, and returned to Babylon; but giving himself over to pride and idolatry, he was driven from his kingdom.

6. **About this time** occurred the events recorded in the Old Testament, concerning Daniel and the fall of Babylon.

7. **Babylon** was taken by Cyrus, king of Persia. It is now desolate.

8. **It is believed by many** that the building of the tower of Babel was begun here.

9. **The Persian Empire** was founded by Cyrus, who united Persia and Media.

10. What two rivers unite and flow into the Persian Gulf? Between those two rivers was once a fertile and populous region, which is now a desert. Mention it.

11. On what river was the ancient city of Babylon? What region lay south of Babylon? What empire was east of Mesopotamia? What cities in Syria? What city at the mouth of the Nile?

Where were Thebes and Memphis, the ancient capitals of Egypt?

What city now stands on the Nile nearly opposite the site of Memphis? *Ans.* Cairo. What region west of Egypt? What parts of Africa were south of the Mediterranean Sea?

12. What was France formerly called? The Strait of Gibraltar? The Bay of Biscay? England? Scotland? Ireland? In what part of Europe was Dacia? Thrace? Sarmatia? Macedonia?

13. **The great empires** of antiquity were Assyria, Babylon, Persia, Greece, and Rome.

14. **Rome** was founded by Romulus more than 700 years B. C., but its early history is not authentic.

15. **Its territory,** for several centuries, was of limited extent; but about the beginning of the Christian era, it ruled over nearly the whole of the world then known to the Romans.

16. **Its greatness continued** until about the 4th century, A. D., when it began to fall.

17. **Rome** was first a kingdom, then a republic, and afterwards an empire.

18. **The most celebrated Roman** was Julius Cæsar. He was noted as a general, statesman, orator, and author. He made important conquests in Europe, Asia, and Africa; but soon after he became dictator, he was assassinated.

19. **Julius Cæsar** was succeeded by Augustus, who became emperor.

20. **During the first century** Rome had successively thirteen emperors, the most noted of whom were Augustus, Tiberius, Nero, Vespasian, Titus, Domitian, and Trajan.

21. **Augustus** promoted peace, literature, and the arts, and by his orders, magnificent temples, aqueducts, canals, and baths were built.

22. **He was emperor of Rome** about forty years, including the first ten years of the Christian era.

23. **The first Christian emperor** was Constantine, who removed the capital from Rome to Byzantium, named from him Constantinople, about 300 A. D.

24. **Ancient Italy:**—Mention its northern divisions,—southern,—central.

Where is the Rubicon River, celebrated on account of its passage by Cæsar, who thus declared war against the republic.

Where were Pompeii (*pom-pay'e*) and Herculaneum, cities destroyed by an eruption of Mt. Vesuvius?

Where was the Sicilian (now Messina) Strait? The whirlpool Charybdis (*ka-rib'dis*)? The rock Scylla?

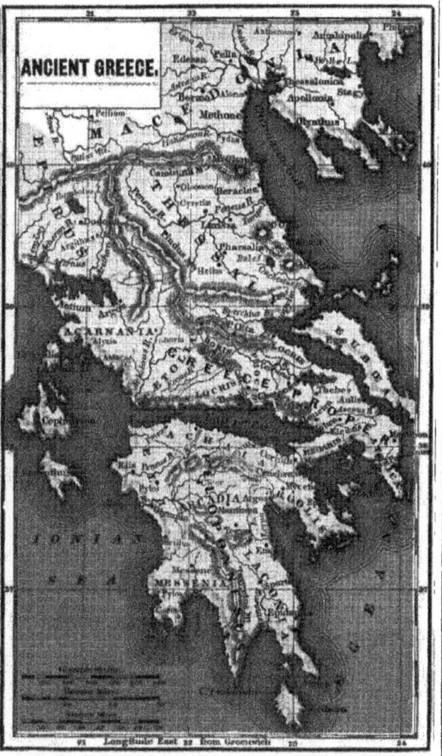

1. **ANCIENT GREECE.** What country was north of Ancient Greece? What sea east? West?

What gulf extends nearly across? What two divisions of Greece in the north?

What mountain on the northern boundary? West of Thessaly?

2. **The most powerful states** of Greece were Athens and Sparta.

Where is Athens? Where was Sparta situated?

3. **Northeast of Athens** is the plain of Marathon, celebrated for the victory of Miltiades over King Darius, successor of Cyrus (B. C. 490).

Where is the pass of Thermopylæ, celebrated for the battle in which Leonidas and 300 Spartans perished in defending Greece against the invasion of Xerxes, the successor of Darius?

4. **Xerxes** destroyed Athens, but as his navy was soon after defeated near Sal'a-mis, he was compelled to return to Persia (B. C. 480).

Where is the island of Salamis?

What ancient town, near Mt. Parnassus, was famous for its oracle of Apollo?

Where was Olympia, celebrated for its temples and the Olympian games?

Where is Corinth, once the richest and most flourishing city in Greece?

5. **Corinth** was destroyed by the Romans (B. C. 146).

6. **The Athenians and Corinthians** were long celebrated for their learning, refinement, and wealth. (See page 75.)

7. **The Greeks** received their knowledge of the arts, etc., originally from the Egyptians and Phœnicians.

8. **The climate** of Southern Europe was formerly not so warm as it is now.

9. **Greece was taken** by Philip of Macedon in the 4th century, B. C. He was succeeded by his son, Alexander the Great, who fought many battles and captured Asia Minor, Phœnicia, Egypt, Persia, Media, and parts of India.

10. **After the death of Alexander the Great,** his empire was divided, Egypt and Palestine passing under the rule of Ptolemy, whose dynasty lasted about 300 years.

11. **The last of this line of rulers** was Cleopatra, queen of Egypt, who committed suicide.

ROUTES OF TRAVEL I

Maine.—What city is the principal railroad center in Maine? Through what city would you pass in going by railroad from Portland to Bangor? What places on the railroad between Portland and Boston? Portland and Montreal? Portland and Quebec? Portland and Newport?

New Hampshire.—What railroad begins at Portland and passes through New Hampshire and Vermont? *Ans. Grand Trunk Railroad.* At what place on that road would you stop to visit Mt. Washington? *Ans. Gorham.* By what routes can you go from Concord to Montreal?

Vermont.—What places on the routes between Lake Champlain and Boston? Burlington and Albany? Burlington and Hartford?

Massachusetts.—Which is the principal railroad center? Through what places would you pass on the Boston and Albany Railroad? Between Boston and New York, by way of Springfield? By way of the Connecticut Shore?

Connecticut.—Which is the principal railroad center? What places on the route between New Haven and Boston? New Haven and Quebec? New Haven and Montreal?

New York.—What railroad runs east and west through the center of the State? Through the southern part? Mention some of the places on the New York Central road. On the Erie road.

What railroad runs along the east bank of the Hudson River? *Ans. Hudson River Railroad.* Mention some of the places on the Hudson River Railroad. On the railroad between Albany and Ogdensburg. Between Watertown and Lake Champlain. Between Albany and Oswego. By what routes can you go from New York to Buffalo and Suspension Bridge?

Pennsylvania.—Mention the principal railroad centers. What city is almost equally distant from Philadelphia, Harrisburg, and Easton? What places on the Pennsylvania Central Railroad? On the road between Harrisburg and Buffalo? Harrisburg and Erie? Harrisburg and Easton?

New Jersey.—What city is about one-third the distance from Philadelphia to New York? What city on the Delaware River opposite Philadelphia? With what places on the coast is Camden connected by railroad?

Delaware.—What large city in Delaware on the route between Philadelphia and Baltimore? By what other route can you go from Philadelphia to Baltimore? On what waters would you sail from Philadelphia to Baltimore? Philadelphia to New York?

Maryland.—Which is the chief city and railroad center in Maryland? In what direction must you go from Baltimore to Philadelphia? To Washington? To Annapolis? To Harrisburg? To Harper's Ferry?

Virginia.—What places on the railroad between Washington and Goldsboro, North Carolina? Between Washington and Knoxville, Tennessee? Norfolk and Lynchburg?

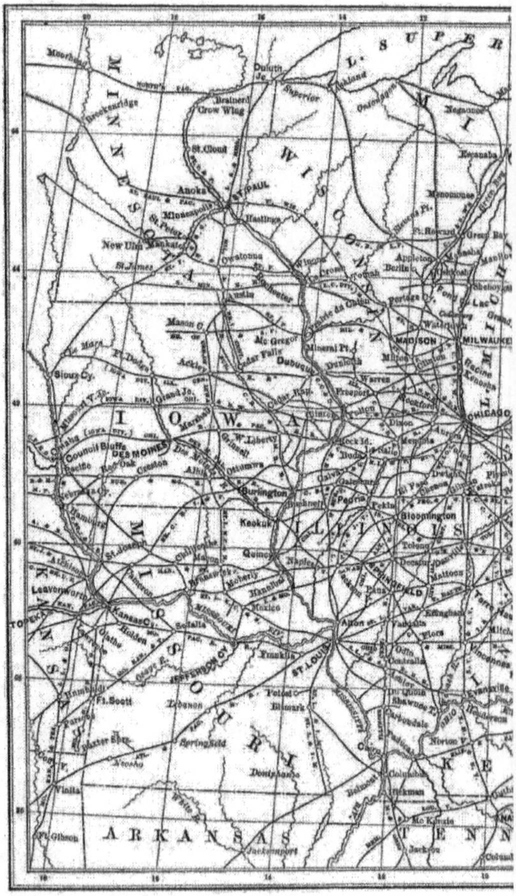

Ohio.—Where is the Lake Shore and Michigan Southern Railroad? Through what places would you pass in going from Cincinnati to Columbus? Columbus to Cleveland? Cincinnati to Sandusky? Cincinnati to Wheeling?

Indiana.—Mention the principal railroad centers. What places on the route from Indianapolis to St. Louis? To Cincinnati? Logansport to Detroit? Fort Wayne to Milwaukee?

Illinois.—Through what places would you pass on your way from Chicago to St. Louis? St. Louis northeast to Terre Haute? Aurora west to Rock Island? Springfield northeast to Logansport, Indiana? Chicago southwest to Quincy?

Michigan.—Mention the different routes from Detroit to Grand Haven. What places on the Michigan Central Railroad? What places on the route from Detroit northwest to Saginaw?

Wisconsin.—Which is the chief city in Wisconsin? What places on the railroad from Milwaukee to Chicago? To La Crosse? To Prairie du Chien?

Minnesota.—What railroads cross the State from east to west? W Places on the railroad which runs southwesterly from St. Paul? Souther Southeasterly?

Iowa.—At what city in Nebraska does the Union Pacific Railroad beg *Ans. Omaha.* What railroad extends from Council Bluffs to Chicago? *Ans. C cago, Rock Island, and Pacific Railroad.* Mention some of the places in Iowa this road,—on the Des Moines Valley Railroad,—on the Illinois Central road,— the Burlington and Missouri River Railroad. Mention some of the places on route from Burlington to St. Paul, Minnesota.

Missouri.—Which is the principal railroad center in the eastern part of State? Northwestern? What railroad runs southwesterly from St. Louis? W erly? What places on that road? What places on the Hannibal and St. Jos Railroad? What places on the northern route between St. Louis and Kansas Ci

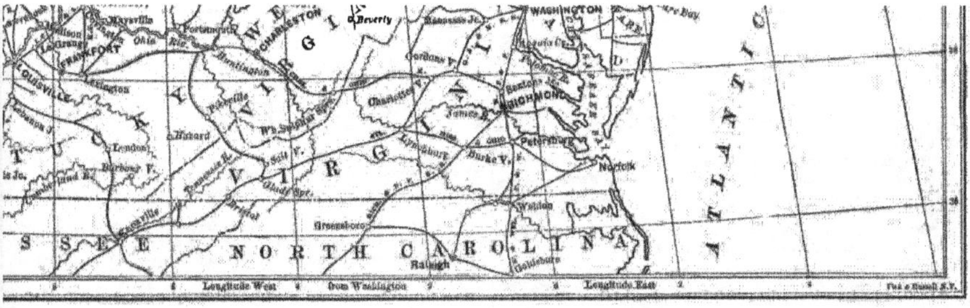

Kentucky.—What places on the route from Louisville to Cincinnati? Louis-le southwest to Paducah? How would you go from Louisville to Frankfort? Nashville?

REVIEW QUESTIONS.

What cities would you pass through on your way from New York to Boston? w York to Montreal? New York to Buffalo? New York to Pittsburg? New rk to Washington? What places would you pass through on your way from ston to Oswego? Boston to Cleveland? Cleveland to Cincinnati? Cleveland Chicago? Chicago to Nashville?

Which is the most important railroad in Canada? How far north and east does extend? What places on the Grand Trunk Railroad are on Lake Ontario? On Lawrence River?

NOTE.—*Steamboats can sail down the St. Lawrence River, but on account of the rapids above or southwest of Montreal, they return by canal.*

By what route would you go from Montreal to Toronto? To St. Paul? *Cross Lake Michigan by steamer.*

With what railroad in Canada does the New York Central connect at Suspension Bridge? Where does the Great Western Railroad of Canada connect with the Grand Trunk Railroad? What is the shortest route from Rochester to Detroit? Which are the two principal routes between New York and Toronto?

By way of what large cities would you go from Detroit to Cincinnati? From Pittsburg to Louisville? From Buffalo to St. Louis? From Kansas City to Cincinnati? From Cincinnati to Philadelphia? From Milwaukee to Nashville? From Philadelphia to White Sulphur Springs, West Virginia? From Chicago to Omaha? From Boston to Norfolk?

A TOUR IN EUROPE.

LEAVING NEW YORK in a steamer, you sail in a southerly direction—through New York Bay and the Narrows, to the Atlantic Ocean; then, sailing in an east northeasterly direction for nine or ten days, you arrive at Brest, the nearest port of France, having sailed the distance of 3000 miles. Here you leave the steamer and begin your tour in Europe.

2. You could have remained in the steamer until its arrival, the next day, at Havre. Steamers of other lines leave New York for Liverpool, Glasgow, Bremen, and Hamburg.

3. If you leave home on a cold day in autumn, you would, on entering Brest be surprised at the mildness and moisture of its climate, and to see flowers blooming in the open air. You would wonder to see its narrow streets and odd-looking houses, and the peculiar dress of the people. Many of the women wear high white caps, instead of hats or bonnets, and as they walk, their loose wooden shoes make a continual clatter.

4. *At Brest* you take the cars for Paris, and ride first through the rough and barren province of Brittany, and afterward through a beautiful and well-cultivated part of France. A ride of sixteen hours brings you to the gayest city in the world.

5. *Going through Paris*, you remark the politeness and animation of the people who crowd its broad boulevards and its numerous cafés and restaurants. You notice also the large handsome buildings, showy store-windows, its small parks and plazas, its fountains, columns, and triumphal arches.

6. *Among the places of interest* which you visit in Paris are the Palace of the Louvre, with its famous museum of paintings, frescoes, sculpture, and antiquities; the Palais Royal, with its garden and shops; the Cathedral of Notre Dame; the Church of the Madeleine; the Bois du Bologne (a park); and the Palace of Versailles, twelve miles from Paris.

7. The Palace of the Tuileries was burned by the Communists in 1871. It was founded more than 300 years ago, and became the residence of the French monarchs. Napoleon III greatly improved it and connected it with the Palace of the Louvre.

8. *As Paris is the great railroad center* of France, you can go by the cars in every direction; north, to Calais or Boulogne, thence by steamboat to England; northeast, to Belgium; east, through Champaigne, celebrated for its fine wines, to Germany; southwest, to Spain, by way of the city of Bordeaux, celebrated for its extensive wine trade; or southeast, to Switzerland and Italy.

9. *Taking a southeasterly direction*, you pass through the famous wine-growing district of Burgundy, where the plains and hills are covered with vines. (See illustration on page 68.)

10. *Crossing the Cote d'Or Mountains*, you ride along the bank of the Saone River to its junction with the Rhone, and arrive at an ancient city which is, next to Paris, the largest city in France, and the chief seat of its silk manufacture. *What city is it?* Here, as in every other large city in Europe which you visit, your attention is directed to its churches, museums, picture galleries, narrow streets, and high buildings.

11. *Silk-weaving* in Lyons is carried on in the dwellings of the workmen. Cotton and woolen goods, jewelry, and bronzes are also extensively manufactured.

12. The population of Lyons is nearly one-third of a million, being about the same as that of St. Louis, the metropolis of Missouri.

13. *As you descend the Valley of the Rhone*, you observe the picturesque towns and villages which line the banks of that river, and the numerous stone bridges which span its rapid current; and, in every direction, you see rich vineyards, besides orchards of olive, mulberry, and other trees.

14. The leaves of the mulberry tree constitute the food of the silkworm. The olive tree yields a valuable oil, and its fruit is extensively used as an article of diet.

15. This section of France is noted for the manufacture of ribbons, velvets, and other silk goods.

16. After riding twenty hours in the cars from Paris, you reach the ancient city of Marseilles (founded 600 B. C.), the chief commercial city of France. Here you learn that the ships which crowd its fine harbor visit all the principal ports of the Mediterranean Sea.

17. Traveling eastwardly along the coast, you observe that the climate in midwinter is as warm and delightful as at home during the month of May. Soon after leaving Marseilles, you pass through Toulon, which is, next to Brest, the principal naval station of France; and, further on, you arrive at Nice (*neece*), which is a popular winter resort for invalids and tourists from every part of Europe and the United States.

18. *Near Nice* is Mentone (*men-to'nay*), situated in the small and ancient principality of Monaco. In the gardens and on the hill-sides all about these places, the orange and lemon trees are loaded with ripe fruit, and flowers are in full bloom even in midwinter.

19. Which is further south, Nice or Chicago? (See page 28.)

20. *From Nice to Gen'oa* you may go either by railroad along the coast (six hours), or by carriage on the Corniche (*korneesh'e*) road (two days, spending the night in one of the villages on the route).

21. *The Corniche Road* is unsurpassed for the grandeur, beauty, and diversity of its scenery. It is in some places high up on the mountain side and on the edge of precipices; in others it leads down to the narrow streets of the towns and villages near the sea. Groves of olive trees cover some of the mountains even to their summits.

22. Here and there, on the tops of hills and crags, you see picturesque villages, ancient towers, and ruined castles; and far below you, the blue waters of the Mediterranean.

23. The shortest route between Paris and Italy is by way of the tunnel through Mt. Cenis. (See page 65, paragraph 1.)

24. *In Genoa* you notice its narrow, uneven, and crooked streets; its high stone houses; grand old palaces; flat roofs, on which are flowering plants, orange and lemon trees; and its manufactures of gold and silver jewelry.

25. *From Genoa to Naples* you may go by steamer, stopping one day at Leghorn, which ranks after Marseilles, Genoa, Trieste, and Smyrna, as a Mediterranean seaport. Leghorn has an important coral fishery, and its manufactures comprise straw hats, and silk, woolen, and cotton goods. Near Leghorn is the city of Pisa, situated, like Florence, on both sides of the River Arno. Here you ascend the celebrated leaning tower, which is 190 feet high and constructed of white marble.

26. Returning to the steamer, you continue your voyage down the coast and reach Naples the following day.

27. Entering the beautiful Bay of Naples, you obtain a fine view of Vesuvius and the chief city of Italy, the latter extending along the shore and up the sides of adjacent mountains.

28. *In Naples* you observe that the streets are very narrow, uneven, and generally without sidewalks; that the houses are high and substantial; that the streets are paved with blocks of lava; and that the people, horses, and donkeys travel in the streets promiscuously.

QUESTIONS.—PARAGRAPH 1.—In what directions would you sail from New York to France? What is the nearest French port? What is the distance between New York and Brest? Would you be sailing with or against the Gulf Stream? In what time?

2. Where is Havre? (*Refer to the maps.*) Liverpool? Glasgow? Bremen? Hamburg?

3. What can you say of the climate of Brest and other parts of Western Europe? (*See page 60, paragraph 10.*) What can you say of the appearance of Brest and of its people?

4. Describe your journey from Brest to Paris.

5. What can you say of the people of Paris? What would you notice especially in riding or walking through the city?

6. Mention some of the interesting places in Paris. What palace twelve miles from Paris?

8. What city is the great railroad center of France? Where would you leave France for England? What celebrated wine district east of Paris? Southeast? Where is the city of Bordeaux situated? For what is it celebrated?

9. What can you say of Burgundy? What mountains and river run through it?

10. How is Lyons situated? What can you say of its size? Its manufactures? Its streets and buildings?

13. What can you say of the Valley of the Rhone?

16. What can you say of the city of Marseilles, its harbor and trade?

17 and 18. What can you say of the climate on the Mediterranean coast, between Marseilles and Italy? What places on that coast? What fruits are found on the trees even in winter?

20. By what routes can you go from Nice to Genoa? What time is required for each?

21 and 22. What can you say of the Corniche road? Its scenery? What trees abound?

23. What is the shortest route from Paris to Italy?

24. What can you say of the streets of Genoa? Its houses? Its manufactures?

25. Sailing from Genoa to Naples, what direction would you take and at what city would you stop? What can you say of Leghorn as a seaport? What fishery has it? What manufactures? Where is the Leaning Tower? What can you say of it? What island do you pass in sailing between Italy and Corsica?

27. Entering the Bay of Naples, what city, volcano, and small islands do you see? What can you say of the situation of Naples?

29. In almost every direction you see in the gardens orange and lemon trees filled with fruit and blossoms even in midwinter, and wonder when you think that Naples, with such a delightful winter climate, is in the same latitude as the city of New York. (See page 72, paragraph 9.)

30. ***Near Naples*** are the famous volcano Vesuvius and the ruined cities of Pompeii and Herculaneum. (See page 73, paragraph 6.)

31. A short ride from Naples brings you to Vesuvius. Ascending the volcano from one of the towns at its base, you pass, for a considerable distance, cottages, vineyards, and gardens; above which, the surface is all rough and rugged with hardened lava and black, burnt stones. Everywhere on its summit you see sulphurous smoke issuing from awful looking openings, and the pieces of lava feel like hot coals under your feet.

32. *Pompeii* was an ancient city of 25,000 inhabitants, beautifully situated near Vesuvius. Although visited by an earthquake in the year 63, its total destruction was not effected until sixteen years afterward, when it was overwhelmed with ashes, cinders, and liquid mud, during an eruption of the volcano. Most of the inhabitants escaped. The city was completely buried by this and subsequent eruptions, and remained so for nearly 1700 years.

33. ***The ruins of Pompeii*** (*pom-pay'e*) are visited by many travelers in Europe. In the portion of the city which has been uncovered you see the ruins of its temples, forums, theaters, and dwellings, with their columns, statuary, mosaics, and frescoes. You notice that its streets are very narrow, most of them but one or two paces from curb to curb; that they are paved with blocks of lava, and much worn by the chariot-wheels.

34. Many fine mosaics, bronzes, and other works of art found in Pompeii and Herculaneum are now in the museum of Naples.

35. ***Herculaneum,*** destroyed at the same time, is on the road between Naples and Pompeii.

36. Some travelers take the steamer at Naples for Palermo, Malta, Athens, Smyrna, Egypt, and the Holy Land. Others return north by the cars to Rome (seven hours).

37. ***Rome*** is situated on both sides of the Tiber, and is surrounded by walls. The larger portion of the city is on the eastern side of the river, and contains the famous ruins of the Roman Forum, Coliseum, the baths of Caracalla, and the Forum of Trajan, which were in their grandeur more than eighteen centuries ago; besides churches of more modern date, which are celebrated for valuable sculptures and paintings.

38. ***On the western side of the river*** is St. Peter's Cathedral, adjoining which is the Vatican, the Pope's palace. The Vatican contains galleries of celebrated sculptures, paintings, etc. Here is Raphael's painting of the Transfiguration, considered the best, or one of the best paintings in the world; here, also, is the celebrated fresco of the Last Judgment, by Michael Angelo, the distinguished architect, sculptor and painter.

39. ***From Rome to Florence*** (ten hours) the route is varied and picturesque. Florence, situated on both sides of the Arno, is celebrated for the beauty of its situation. It contains handsome buildings, gardens, squares, fountains, statues, bridges, etc., and is surrounded by hills which are ornamented with villas, groves, and gardens. Among its numerous works of art is the celebrated statue of the Venus de Medici (*med'e-che*).

40. ***Traveling from Florence to Venice,*** you cross the Apennines and the alluvial plain or valley of the Po, passing through the cities of Bologna, Ferrara, and Padua.

41. ***Venice*** is unlike every other city which you have visited, for it has canals instead of streets, and gon'dolas instead of carts and carriages. Its chief attractions are the Grand Canal and the square of St. Mark. The Grand Canal is lined with magnificent old palaces and other buildings.

42. ***The Venetian manufactures*** include glass, mirrors, jewelry, beads, and artificial pearls.

43. The city covers seventy-two islands, which are connected with each other by numerous stone bridges. The most noted of these bridges are the Rialto, which spans the Grand Canal, and the "Bridge of Sighs," across which prisoners passed from the palace to the prison for sentence.

44. ***On your way from Venice to Vienna,*** you stop at Trieste, the chief commercial city of Austria; then over mountains remarkable for numerous tunnels and viaducts, and across valleys containing well-cultivated farms and vineyards.

45. ***Vienna*** is one of the handsomest and wealthiest cities in Europe. It contains several palaces, and numerous churches, squares, gardens, fountains, and picture galleries. The people are especially fond of music. The Prater—a park or wood—is a famous resort on pleasant afternoons, and its long avenue, lined with restaurants, etc., is usually crowded with equipages.

46. ***A favorite excursion*** is to Schonbrunn, where the Emperor has his summer palace. Its gardens are a marvel of beauty; their long walks or avenues are bordered with gigantic hedges of forest trees, so clipped as to resemble great walls, in which, at intervals, are niches filled with statuary.

47. ***Among the manufactures*** in which Vienna excels are articles in Russian leather, porcelain, velvet, and silk, besides bronzes, meerschaum pipes, and Bohemian glass.

48. ***From Vienna to the city of Dresden*** your course is northwest, through the ancient kingdom of Bohemia, a fertile and beautiful plain inclosed by mountains. It is now under the dominion of Austria.

49. Stopping at Prague, its capital, you are struck with the quaint appearance of its narrow streets and the architecture of its old buildings.

50. ***You reach Dresden*** after a ride through a region remarkable for picturesque scenery—over mountains and valleys, past crags with old castles or fortresses on their tops and small villages at their base.

51. ***Dresden*** is situated in the rich valley of the Elbe, which contains beautiful fields, vineyards, groves, gardens, and orchards.

52. ***The chief attractions*** in Dresden are its picture-gallery and the "Green Vaults." The former contains, among other celebrated paintings, that of the Madonna di San Sisto, by Raphael; the "Green Vaults" are a number of rooms where are exhibited jewels, crowns, scepters, etc., famous for their value and history.

53. ***From Dresden north to Berlin,*** the capital of the German Empire, the route is generally level.

54. ***Berlin*** is a beautiful city, on both sides of the Spree River. Its chief attractions are its splendid palaces and other buildings on *Unter den Linden*, its principal street, its numerous institutions of learning, its parks, gardens, and statues. The people are noted for their love of music. Berlin is an important manufacturing city. Near Berlin is Potsdam, celebrated for its palaces, beautiful gardens, and villas.

55. ***From Berlin*** you may go west to Hanover; thence northwest to the free city of Bremen, which is, next to Hamburg, the chief commercial city in Germany, and take the steamer for New York; or, you may leave Hanover in a southwesterly direction for Cologne, which is celebrated for its magnificent cathedral; thence to Coblentz, where you take the steamboat on the Rhine for Mayence. (See page 69.)

56. From Mayence you go to Frankfort-on-the-Main, which is one of the oldest and wealthiest cities in Germany. Here the German Emperors were crowned; and on these occasions, they were waited upon by kings.

57. ***Near Frankfort*** are the celebrated watering-places Homburg, Wisbaden, and Ems.

58. From Frankfort south, by railroad, you pass through Darmstadt and the Grand Duchy of Baden, and arrive at Switzerland.

28-30. What can you say of the streets of Naples? Its buildings? Its people, etc.? Its climate? What are seen in its gardens in winter? What city in the United States is in the same latitude as Naples? Do fruit and flowers grow in the open air in New York, during the winter? What three celebrated places near Naples?

31, 32. What would you see in making the ascent of Mt. Vesuvius? What can you say of its summit? How was Pompeii destroyed?

33, 34, and 35. What can you say of Pompeii as it appears now? Where are many of its works of art? Where are the ruins of Herculaneum?

36. Travelers bound southward and eastward take steamers at Naples for what places? Where is Palermo? Malta? Athens? Smyrna? Egypt? The Holy Land?

37, 38. In what direction is Rome from Naples? What celebrated cathedral in Rome? What can you say of the Vatican? Mention the principal ruins.

39-43. What can you say of the route between Rome and Florence? How is Florence situated? How surrounded? What plain and cities do you pass through on your way from Florence to Venice? What can you say of Venice?

44. Describe the route from Venice to Vienna?

45-47. What can you say of Vienna? The Prater? Schonbrunn? What are manufactured in Vienna?

48-50. What is your course from Vienna to Dresden? What can you say of Bohemia? Prague? What can you say of the scenery between Prague and Dresden?

51-53. How is Dresden situated? What are its chief attractions? What is the route between Dresden and Berlin?

54. What can you say of Berlin? Its situation? Into what does the Spree empty? What is the population of Berlin? (See page 69, paragraph 14.) Its latitude? What is the latitude of Quebec? Of the northern boundary of Minnesota? What can you say of Potsdam?

55. What is the chief commercial city in Germany? What city is the second in commercial importance? Where do most German emigrants take the steamer for America? Where is Cologne situated? Coblentz? Describe the scenery on the Rhine. Where is the best scenery?

56. Where do you leave the steamboat in order to visit Frankfort? What can you say of Frankfort? What imposing ceremonies took place here?

57, 58. What celebrated watering-places near Frankfort? Through what places would you pass on your way from Frankfort to Switzerland?

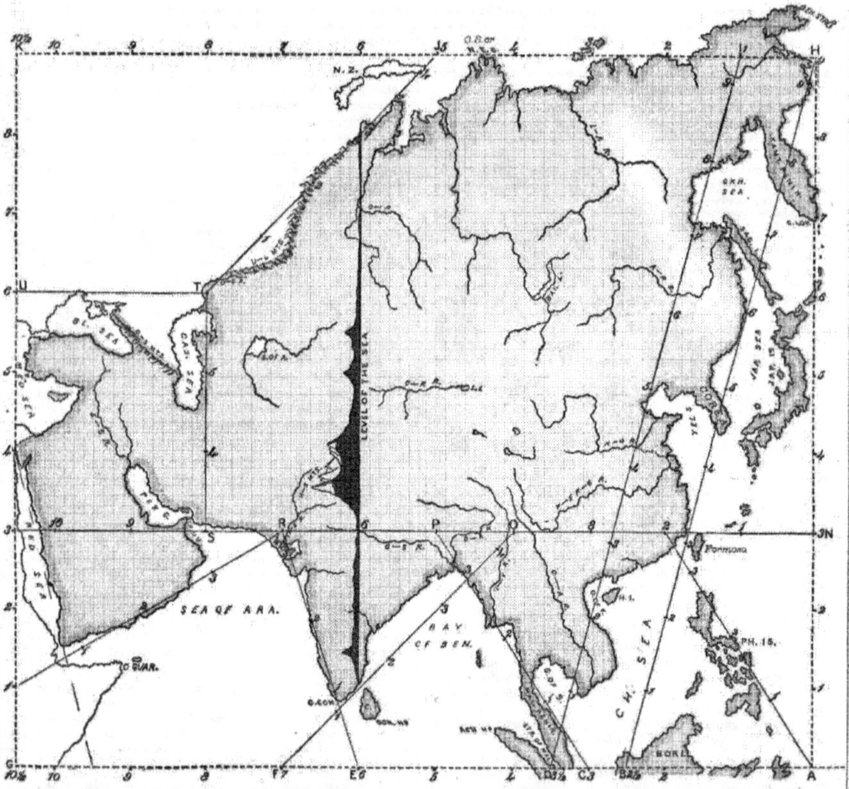

DIRECTIONS FOR DRAWING ASIA.

When you draw any place write or print its name in full, if the space will allow it. If not, then its initials. (See map on p. 76.)

1. Begin at A and mark the point H 9 ms. north, and G 10½ ms. west. Mark the points 1 m., 2 ms. and 2½ ms. west of A, and draw ***Borneo.*** From B measure towards H 3 ms., and draw ***Formosa*** and a part of the coast of ***China.*** Next mark 5 ms. and draw ***Corea;*** mark 6 ms. and draw the coast of ***Mantchooria;*** mark 7 ms. and draw ***Saghalien Is.;*** mark 8 ms. and draw ***Kamtschatka,*** and 9 ms., and draw the coast to ***Behring Strait.***

2. Mark the point D 3½ ms. west of A, and I 1 m. west of H. North of D, towards I, mark 1 m., 2 ms., 3 ms., and draw ***Sumatra,*** the ***Malay Peninsula,*** the ***Gulf of Tonquin*** and ***Hainan Island.*** Mark 5 ms. and draw the ***Yellow Sea,*** 7 ms. and 8 ms., and draw the ***Sea of Okhotsk.*** Southeast of ***Saghalien Is.*** and ***Corea*** draw the ***Japan Islands.*** Between ***Formosa*** and A, draw the ***Philippine Islands.***

3. Connect 3 ms. north of A with 3 ms. north of G, and mark the measurement points O, P, and R. Between A and G mark the points C, D, E, and F. Mark the measurement points from C to P, and draw the northeast coast of the ***Bay of Bengal.*** Mark the points from F to O, and complete the coast line to ***Cape Comorin;*** then draw ***Ceylon.*** Mark the points from E to R, and draw the east coast of the ***Arabian Sea,*** locating the ***Gulfs of Cambay*** and ***Cutch.*** Mark S 1 m. west of R, and the points from 1 m. north of G to R, and draw the northern coast of the ***Arabian Sea,*** the ***Strait of Ormus,*** the ***Persian Gulf,*** the coast of ***Arabia,*** the ***Strait of Babel Mandeb, Cape Guardafui*** (*fwee'*), and a part of the coast of ***Africa.***

4. Mark the point 4 ms. north of G, and draw the ***Red Sea.*** Mark 6 ms. at U, and T 3 ms. north of S, and draw the eastern part of the ***Mediterranean Sea,*** the ***Black Sea,*** the ***Caspian Sea,*** and the ***Caucasus Mountains.***

5. Mark the points west of H to J, and the points from T to J, and draw the ***Ural Mts., Gulf of Obi, Nova Zembla, Sea of Kara, North-East Cape,*** and the north coast line to ***Behring Strait.***

6. Draw the ***Mountains.***

7. Draw the ***Rivers,*** commencing at the source of each.

8. Indicate by dotted lines the boundaries of the ***Countries.***

9. Mark all the ***Peninsulas, Islands, Seas, Gulfs, Bays, Capes, Cities,*** and ***Towns.***

The Grand Divisions of the Earth.

Grand Divisions.	Area.	Population.
Asia	16,415,758	751,062,475
Africa	11,556,660	191,000,000
North America	9,059,927	59,539,474
South America	6,954,131	27,562,600
Oceanica	4,388,025	30,248,333
Europe	3,830,857	301,752,419
Total	52,204,848	1,361,196,351

Countries, Where situated?	Area in Sq. Mi.	Populations.	Government.
Austria	240,319	36,000,000	*Empire*
Argentine Republic	879,800	1,812,000	*Republic*
Afghanistan	258,500	4,000,000	—
Andorra	149	12,000	*Republic*
Arabia	1,026,640	4,000,000	—
Balize	14,000	10,000	—
Brazil	3,230,000	11,780,000	*Empire*
Belgium	11,318	5,087,105	*Kingdom*
Bavaria	29,373	4,861,402	*Kingdom*
Baden	5,712	1,461,428	*G. Duchy*
Bolivia	535,000	1,980,000	*Republic*
Chinese Empire	4,700,000	446,000,000	*Empire*
Chili	250,000	2,000,000	*Republic*
Colombia, U. S. of	357,000	3,000,000	*Republic*
Costa Rica	21,500	185,000	*Republic*
Denmark	14,784	1,784,741	*Kingdom*
Ecuador	218,000	1,371,000	*Republic*
France, inc. Corsica	204,079	36,102,821	*Republic*
Gt. Britain and Irel'd	122,550	31,817,108	*Kingdom*
Germany	210,778	41,058,196	*Empire*
Greece	19,341	1,457,894	*Kingdom*
Guatemala	44,780	1,180,000	*Republic*
Hindoostan	1,500,000	180,500,000	—
Holland, inc. Lux'g	13,890	3,915,966	*Kingdom*
Honduras	47,092	350,000	*Republic*
Italy (inc. Islands)	114,396	26,789,038	*Kingdom*
Japan	149,397	33,000,000	*Empire*
Mexico	712,850	9,000,000	*Republic*
Monaco	6	1,877	*Princip'y*
Nicaragua	58,169	400,000	*Republic*
Prussia	135,806	24,693,066	*Kingdom*
Persia	520,000	5,000,000	*Kingdom*
Paraguay	63,000	1,020,000	*Republic*
Patagonia	210,000	120,000	*No Gov't*
Peru	503,000	2,500,000	*Republic*
Portugal	36,510	3,986,153	*Kingdom*
Russian Empire	8,012,955	82,159,630	*Empire*
San Marino	24	7,303	*Republic*
Sweden and Norway	292,000	5,957,107	*Kingdom*
Spain	195,000	16,565,000	"
Switzerland	15,722	2,669,147	*Republic*
Saxony	6,777	2,556,244	*Kingdom*
Siam	300,000	6,300,000	*Kingdom*
San Salvador	7,335	600,000	*Republic*
Turkey in Europe & Turkey in Asia	828,308	31,550,000	*Empire*
United States	2,446,187	38,925,598	*Republic*
Uruguay	66,000	350,000	*Republic*
Venezuela	368,000	2,200,000	*Republic*
Wurtemberg	7,532	1,818,541	*Kingdom*

POPULATION OF LARGEST CITIES IN AMERICA AND EUROPE.

1. London	Where is it?	On what water?	3,252,000
2. Paris	"		1,825,000
3. Constantinople	"		1,075,000
4. New York	"		1,249,898
5. Berlin	"		825,000
6. Philadelphia	"		674,000
7. St. Petersburg	"		667,000
8. Vienna	"		627,000
9. Liverpool	"		493,000
10. Manchester	"		480,000
11. Glasgow	"		477,000
12. Naples	"		449,000
13. Rio Janeiro	"		420,000
14. Moscow	"		399,000
15. Brooklyn	"		485,252
16. Birmingham	"		343,000
17. Madrid	"		332,000
18. Lyons	"		324,000
19. St. Louis (Mo.)	"		*414,393
20. Marseilles	"		300,000
21. Chicago	"		*325,498
22. Baltimore	"		267,000
23. Amsterdam	"		265,000
24. Leeds	"		259,000
25. Warsaw	"		252,000
26. Boston	"		341,919
27. Dublin	"		246,000
28. Rome (Italy)	"		244,000
29. Sheffield	"		240,000
30. Hamburg	"		240,000
31. Lisbon	"		224,000
32. Palermo	"		219,000
33. Cincinnati	"		*266,000
34. Turin	"		208,000
35. Breslau	"		208,000
36. Havana	"		205,000
37. Pesth	"		202,000
38. Milan	"		199,000
39. Edinburgh	"		196,000
40. Bordeaux	"		194,000
41. New Orleans	"		*203,439
42. Bristol (Eng.)	"		182,000
43. Barcelona	"		180,000
44. Dresden	"		177,000
45. Belfast	"		174,000
46. Brussels	"		171,000

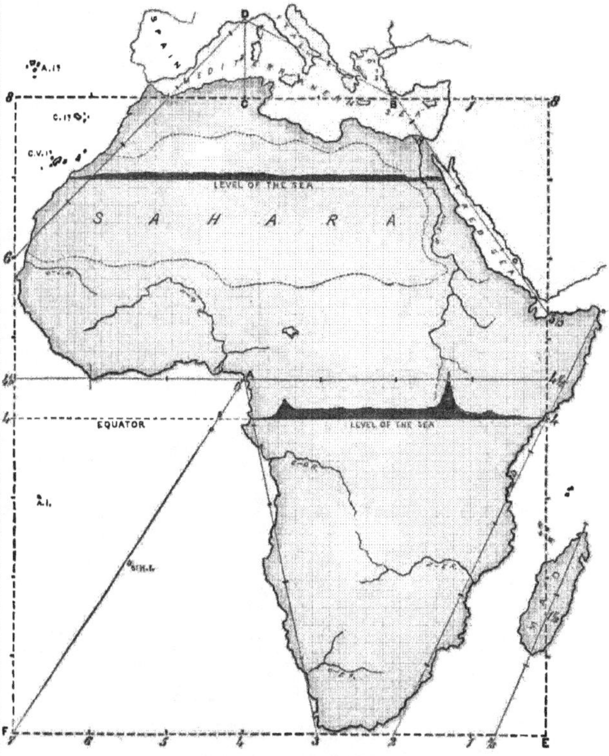

DIRECTIONS FOR DRAWING AFRICA.

Begin at **E**, measure north *8* ms., marking the points *1½*, *4*, *4½*, and *5⅓*. From **E** measure west to **F**, *7* ms., marking the points *⅔*, *2*, and *3*. From *⅔* on this line, through *1½*, north of **E**, measure *3* ms., and draw the island of ***Madagascar.*** From *2* ms. west of **E**, through *4* north of **E**, measure *6* ms. to ***Cape Guardafui***, and draw the south-eastern coast line. From *4½* north of **E**, measure west *7* ms., and mark *4* ms. *5* ms. and *6* ms. From *3* west of **E**, at ***Cape of Good Hope***, measure to **A**, marking the measurement, distances, and draw the coast line.

From *8* north of **E** measure west *8* ms., marking *2* ms. at **B**, and *4* ms. at **C**. At *1* m. north of **C** mark the point **D**. Measure from *5⅓* north of **E** to **B**, and thence from **B** to **D**, and draw the ***Red Sea*** and the eastern part of the ***Mediterranean Sea.*** Connect *6* ms. north of **F** with **D**, mark the measurement distances, and complete the coast lines of the Atlantic and the Mediterranean.

Then from the Map on page 82 mark the capes with their names; also the islands, the mountains, the large lakes; the principal rivers; the Great Desert; next, the countries which border on the Mediterranean Sea, on the Red Sea, on the Indian Ocean and Mozambique Channel, on the Atlantic Ocean and the Gulf of Guinea. Locate the capital of each country.

47. Munich	Where is it?	On what water?	169,000
48. Florence	"		167,000
49. Copenhagen	"		162,000
50. Prague	"		157,000
51. Lille (France)	"		155,000
52. Bahia	"		152,000
53. Adrianople	"		150,000
54. San Francisco	"		149,000
55. Bradford (Eng.)	"		145,000
56. Stockholm	"		138,000
57. Genoa	Where is it?	On what water?	130,000
58. Cologne	"		129,000
59. Venice	"		129,000
60. Newcastle	"		128,000
61. Toulouse	"		127,000
62. Antwerp	"		126,000
63. Lima	"		121,000
64. Buenos Ayres	"		178,000
65. Odessa	"		121,000
66. Hull (Eng.)	"		121,000
67. Dundee	Where is it?	On what water?	119,000
68. Buffalo	"		*124,000
69. Seville	"		118,000
70. Bologna	"		116,000
71. Rotterdam	"		116,000
72. Santiago	"		115,000
73. Nantes	"		112,000
74. Messina	"		112,000
75. Leipsic	"		107,000
76. Montreal	"		107,000

COMPARATIVE SIZES.

Each oblong frame here represents the State of Kansas—200 miles wide by 400 miles long. The states, countries, etc., being drawn on the same scale as Kansas, their sizes are easily remembered.

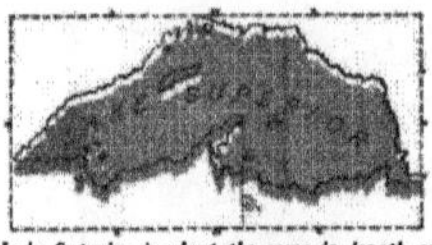

Lake Superior is about the same in length as Kansas—400 miles.

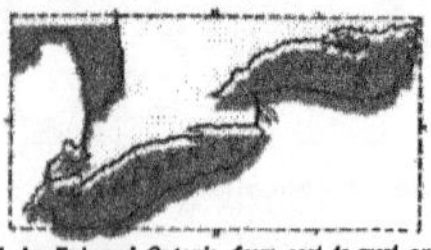

Lakes Erie and Ontario, from east to west, are together nearly 400 miles.

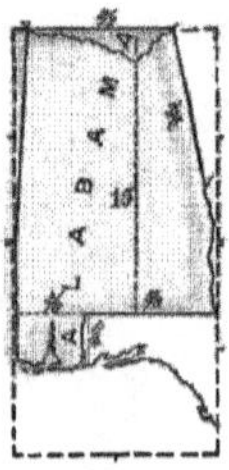

Alabama and Kansas are the same in width—200 miles.

Greece and Kansas are the same in width; Kansas is larger than Greece, Sicily, and the kingdom of Saxony combined.

Florida and Kansas are about the same in length.

Iceland, Palestine, and Kansas are the same in extent from north to south—200 miles.

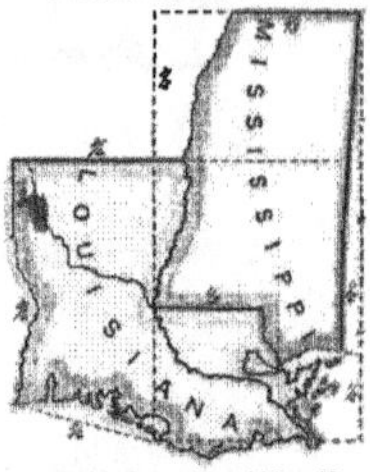

Mississippi is the same in width as Kansas; and the distance from the northern boundary of Mississippi to the southern boundary of Louisiana is equal to the length of Kansas—400 miles.

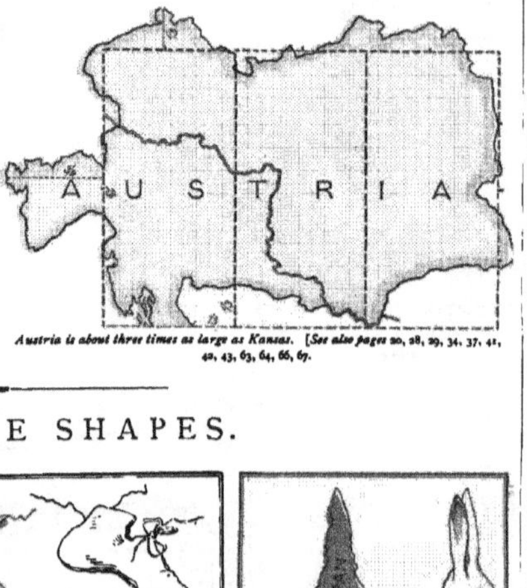

Austria is about three times as large as Kansas. [*See also pages 20, 28, 29, 34, 37, 41, 42, 43, 63, 64, 66, 67.*

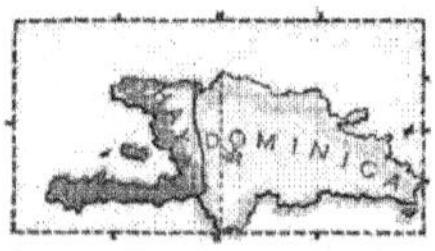

The Island of Hayti is nearly as long as Kansas.

COMPARATIVE SHAPES.

Lake Erie. A whale.

Lake Ontario. A seal.

Black Sea. A child's sock.

Sea of Japan. A Rabbit.

Cuba. A lizard.

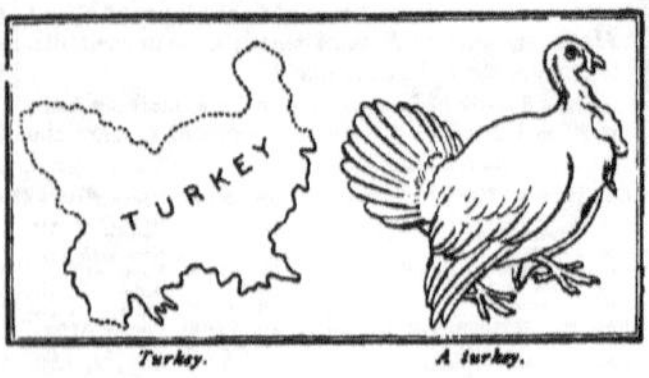

Turkey. A turkey.

New Guinea. A Guinea hen.

PRONUNCIATION OF GEOGRAPHICAL NAMES.

A.

Aar, ahr.
Abaco, ah′bah-ko.
Ab-bi-tib′be.
Abomey, ab-o-may′.
Aboukir, ah-boo-keer′.
Ab-se′cum.
Abydos, ah-bi′dos.
Acapulco, ah-kah-pool′ko.
Ac-co-mac′.
Acheen, atch-een′.
Aconcagua, ah-kon-kah′gwah.
Acquia, a-kwi′a.
Ac′ra.
Acre, ah′ker.
Adalia, ah-dah′le-ah.
A-del′.
A′den.
Ad′ige, or ah′de-ja.
Ad′i-ron-dack.
Ad′riatic.
Ægean, e-je′an.
Afghanistan′.
Agua Dulce, ah′gwah-dool′say.
Aguilas, ah-gool′yahs.
Aisne, ain.
Aix-la-Chapelle, aiks-lah-shah-pel′.
Ajaccio, ah-yaht′cho.
Alamo, ah′lah-mo.
Aland, ah′land.
Albans, St., sent-awl′bons.
Al-be-marle′.
Albuquerque, ahl-boo-ker′kay.
Aleutian, a-lu′she-an.
Algeziras, al-je-zee′ras.
Altai, ahl-ti′.
Altamahaw, awl-ta-ma-haw′.
Alton, ahl′ton.
Altona, ahl′to-nah.
Alvarado, ahl-vah-rah′do.
Am′azon.
Amboy′.
Amherst, ahm′erst.
Amiens, am′i-ens.
Amoo, ah-moo′.
Amoor, or Amur, ah-moor′.
A-moy′.
Amsterdam′.
Anadir, ah-nah-deer′.
Anahuac, ah-nah-wahk′
Anam′, or an′nahm.
Andaman′.
Andes, an′deez.
Andorra, ahn-dor′rah.
An′dover.
Anglesea, ang′gle-se.
Anticos′ti.
Antietam, an-te′tam.
Antigua, an-te′ga.
Antilles, ahn-teel′.
Antisana, ahn-te-sah′nah.
Antuco, ahn-too′ko.
Ant′werp.
Apaches, ah-pah′chayz.
Ap′en-nines.
Appalach′ee.
Appalachico′la.
Apurimac, ah-poo-re-mahk′.
Ar-a-can′.
Ar′al.
Ar′a-rat.
Archangel, ark-an′gel.
Archipelago, ark-e-pel′ah-go.
Ardennes, ar′den, or ar-den′.
Arequipa, ah-re-ke′pah.
Argentine, ahr′gen-tine.
Arica, ah-re′kah.
Arkan′sas.
Armagh, ar-mah′.
Arnheim, ahrn′hime.
Aroos′took.
Arrap′a-hoe.
Artois, ahr-twah.
Ashan′tee, or ash-an-te′.
Asia, a′she-a.
As-sam′.
As-sin′ni-boin.
Astrakhan, ahs-trah-kahn′.
Atacama, ah-tah-kah′mah.
Atchafalaya, atch-af-a-li′a.
Ath-a-bas′ca.
Ath′ens.
At-lan′ta.
Aubigny, o-been-ye′.
Augustine, St., sent-aw′gus-teen.
Australia, aws-tray′le-a.
Auvergne, o-vârne′.
Aux Cayes, o-kay′.
Avignon, ah-veen-yông′.
Avon (Eng.), a′von.
Avon (Fr.), ah-von.
Ayr, air.
As′ov, As′of, or Az′oph.
Azores′, or a-zo′res.

B.

Baalbec, or Balbec, bahl-bek′.
Bab-el-man′deb.
Badajos, bad-a-hose′.
Bad′en, or bah′den.
Bahia, bah-e′a.
Baikal, bi′kahl.
Balaklava, bal-a-klah′vah
Bal-e-ar′ic.
Balize, bah-leez′.
Balkan, bahl-kahn′.
Bal-mor′al.
Bal-zar, bahl-zahr′.
Bankok′, or Bangkok′.
Barba′does, or Barba′dos.
Bar-ne-gat′.
Barn′sta-ble
Basle, bahl.
Bath′urst.
Baton Rouge, bat′on roozh.
Bayonne, bah-yon′.
Bayou, bi′oo.
Beaufort (Br. Dominions), bo′furt.
Beaufort (S. C.), bu′furt.
Beaufort (France), bo-for′.
Bedouin, bed′o-in.
Beersheba, be-er′she-ba.
Behrings, bear′ings.
Be-le-fon-taine′.
Belfast′ (Ireland); Bel′fast (Maine).
Bel-grade′.
Beloochistan′.
Benares, ben-ah′rez.
Bengal, ben-gawl′.
Benguela, ben-gah′la.
Benin, ben-een′.
Ben Lo′mond.
Ber′bera.
Bermudas, ber-moo′das.
Bexar, ba-hahr′.
Beyrout, bay′root.
Biafra, be-af′rah.
Bilbao, bil-bah′o.
Bing′en.
Bistineau, bis-te-no′.
Blenheim, blen′im.
Bogota, bo-go-tah′.
Bokhara, bo-kah′rah.
Bologna, bo-lone′yah.
Bomarsund, bo′mar-soond.
Bombay′.
Bonifacio, bo-ne-fah′tcho.
Bon Homme, bo-nom′.
Bonita, bo-ne′tah.
Boo-tan′.
Boothia, boo-the′ah.
Bordeaux, bor-do′.
Borgne, born.
Borodino, bor-o-de′no.
Bos′po-rus, or Bos′pho-rus.
Boulogne, boo-lone′.
Bourbon, boor′bon.
Bowdoin, bo′den.
Bowling Green, bo′ling.
Brah-ma-poo′tra.
Bras d'Or, brah-dore′.
Brazh′car.
Brazil, brah-zeel′.
Brazos, brah′zos.
Bremen, or bray′man.
Breton, brit′on.
Bucharest, boo-kah-rest′.
Buenaventura, bway′nah-ven-too′rah.
Buena Vista, bway′nah-vis′tah.
Buenos Ayres, bo′nos a′riz.
Bulgaria, bool-gay′re-ah.
Bur′gun-dy.
Bus′so-rah.

C.

Ca-bool′.
Cadiz, kay′diz.
Caen, kon.
Caermarthen, ker-mar′then.
Caernarvon, ker-nar′von.
Cagliari, kahl′yah-re.
Cahawba, ka-haw′bah.
Caicos, ki′koss.
Cairo (Egypt), ki′ro; (U. S.), kay′ro.
Calabar, kah-lah-bar′.
Calais, kal′is.
Calcasieu, kahl′ka-shu.
Callao, kahl-lah′o, or kahl-yah′o.
Cambridge, kame′brij.
Campagna, kahm-pahn′yah.
Canajoharie, kan-a-jo-har′re.
Canandaigua, kan-an-da′gwa.
Can-av′er-al.
Candahar′.
Can-ton′ (China); Can′ton (U. S.)
Cape Girardeau, je-rar-do′.
Cape Haytien, hay′te-en.
Cap′u-a, or kah′poo-ah.
Caqueta, kah-kay′tah.
Caracas, kah-rah′kas.
Car′de-nas, or kar′day-nas.
Caribbe′an.
Car′ibbee.
Carlscrona, karls-kroo′na.
Carlsruhe, karls′roo.
Carthagena, kar-ta-je′nah.
Cashgar′.
Cashmere′.
Castiglione, kahs-teel-yo′nay.
Catoche, kah-to-chay′.
Cat′te-gat.
Cau′casus.
Cayenne, ki-en
Cayman, ki-man′.
Celebes, sel′e-beez.
Cerigo, cher′e-go.
Cevennes, sa-ven′.
Cey′lon.
Chagres, shah′gres.
Chaleur, shah-loor′.
Chapultepec, chah-pool-ta-pek′.
Chateaugay, shat′o-gay.
Chaudiere, sho-de-air′.
Chaumont, sho-mông′, or sho-mo′.
Chatauqua, sha-taw′gwa.
Chemung, she-mung′.
Cherbourg, sher′burg.
Che-sun′cook.
Cheviot, shiv′e-ut.
Cheyenne, she-en′.
Chicago, she-kaw′go.
Chick-a-mau′ga.
Chick′a-saw.
Chicopee, chik-o-pe′.
Chihuahua, che-wah′wah.
Chili, chil′le.
Chil-li-coth′e.
Chiloe, che-lo-ay′.
Chimborazo, chim-bo-rah′zo.
Chincha, chin′chah.
Chinchilla, chin-cheel′yah.
Chincoteague, chin-ko-teeg′.
Chiriqui, che-re-ke′.
Choco, cho′ko.
Chowan, cho-wawn′.
Chuquisaca, choo-ke-sah′kah.
Cienfuegos, se-en-fway′gos.
Cincinnati, sin-sin-nah′te.
Ciudad Real, se-oo-dad′ ra-ahl′.
Civita Vecchia, che′ve-tah vek′ke-ah.
Cler-mont′.
Coahuila, ko-ah-we′lah.
Coatzacoalcos, ko-aht-sah-kwahl′koce
Cobija, ko-be′hah.
Cognac, kon-yak′.
Cohahuila, ko-ah-we′lah.
Cohoes′.
Cole-raine′.
Colima, ko-le′mah.
Cologne, ko-lone′.
Colorado, kol-o-rah′do.
Comayagua, ko-mi-ah′gwah.
Com′o-rin.
Com′o-ro.
Concha, kon′chah.
Concord, konk′urd.
Congaree, kong-ga-re′.
Con-hoc′ton.
Connaught, kon′nawt.
Connecticut, kon-net′e-kut.
Contreras, kon-tray′ras.
Coquimbo, ko-keem′bo.
Coralline, kor′al-lin.
Cor′do-va.
Cor-fu′, or kor-foo′.
Cor′inth.
Cor′ri-en′tes.
Coteau des Prairies, ko-to′day-pray-re′
Cotopaxi, ko-to-paks′e.
Coupee, koo-pay′.
Coxsackie, kuk-saw′ke.
Cracow, kray′ko.
Croatia, kro-a′she-a.
Croix, St., kroi.
Cuenca, kwen′kah.
Cumana, koo-mah-nah
Curaçoa, ku-ra-so′.
Cuyahoga, ki-a-ho′ga.
Cuzco, koos′ko.
Cy-re′ne.

D.

Dahomey, dah-ho′may.
Dalhousie, dal-hoo′ze.
Dalmatia, dal-may′she-ah.
Dalton, dawl′ton.
Dantzic, dant′sik.
Dan′ube.
Dauphin, daw′fin.
Del Fuerte, del fwer′tay.
Del-hi′, or del′le.
Dem′be-a.
Demerara, dem-e-rah′rah.
Derne, der′neh.
Dor′er-et.
Des Moines, de-moin′.
Des Plaines, day-plane′.
Dhawalaghiri, dah-wah-law-gher′re.
Diarbekir, de-ar-bay-keer′.
Dnieper, ne′per.
Dniester, nees′ter.
Dominica, dom-e-ne′ka.
Donegal, don-e-gawl′.
Dongola, dong′go-la.
Dor′chester.
Dordogne, dor-done′.
Drogheda, droh′he-da.
Dubuque, du-bûke′.
Duero, doo-ây′ro.
Dun-dalk′.
Dun′kirk.
D'Urban, dur′ban.

E.

Echelles, Les, lays-a-shel′.
Ecuador′, or Equador′
Edfou, ed′foo.
Edinburgh, ed′in-bur-rah.
Ed′isto.
Elder, I′der.
Elleure′.
El Dorado, do-rah′do.
Eleuthera, e-loo′the-rah.
Elgin, el′ghin.
El Paso del Norte, el-pah′so del nor′ta.
England, ing′land.
Er′e-bus.
E′rie.
Esc′pus.
Erzgebirge, erts-ga-beer′ga.
Espiritu Santo, es-pe′re-to-san′to
Esquimaux, es′ke-mo.
Etienne, ay-te-en′.
Et′o-wah.
Euphrates, u-fray′teez.
Eureka, yoo-re′kah.
Eutaw, yoo′taw.
Euxine, yux′in.
Ev′er-est.
Exuma, ex-oo′mah.

F.

Fauborg, fo′borg.
Falkland, fawk′land.
Falmouth, fal′muth.
Farne, fay′ro.
Fauquier, faw-keer′
Fayal, fi-awl′.
Fay′etteville.
Fee′jee.
Ferrol, fer-role′.
Fez-zan′.
Fichtelgebirge, fik-tel-ga-beer′ga.
Finisterre, fin-is-tare′.
Fiord, fe-ord′.
Flamborough, flam′bur-ruh.
Flo′res.
Foggia, fod′jah.
Fond du Lac, fond-du-lak′.
Formentera, for-men-tay′rah.
Foulah, foo′lah.
Frio, fre′o.
Fron-te-nac′.
Funchal, foon-shal′.

G.

Gaeta, gah-ay′tah.
Galilee, gal′e-le.
Gallinas, gal-le′nas.
Galway, gawl′way.
Gan′ges.
Garonne, gah-rone′.
Gaspe, gahs-pay′.
Gatineau, gah-te-no′.
Gen′o-a.
Geral, zhay-rahl′.
Geysers, ghi′zerz.
Ghent, g hard.
Gihon, je-hon′.
Gila, he′lah.
Gil-bo′a, g hard.
Gil′e-ad.
Glasgow, glas′go.
Gloucester, glos′ter.
Goderich, god′rik.
Gotha, go′tah.
Gothard, got′hart.
Gracias-a-Dios, grah′se-ahs-ah-de-os′.
Granada, grah-nah′dah.
Greenwich, grin′ij, or green′ich.
Guadalaxara, gwah-dah-lah-hah′rah.
Guadalquiv′er.
Guanahani, gwah-nah-hah′ne.
Guanaxuato, gwah-nah-hwah′to.
Guardafui, gwar-dah-fwee′.
Guayaquil, gwi-ah-keel′.
Guaymas, gwi′mas.
Guernsey, ghern′ze.
Guiana, ge-ah′na.
Guyandotte, gi-an-dot′.

H.

Hague, haig.
Hainan, hi-nahn′.
Hakodate, hah-ko-dah′te.
Haussa, how′sah.
Haverhill (Mass.), hay′ver-il.
Haverhill (Eng.), hav′er-il.
Havre de Grace, hav′er-de-grass.
Hawaii, hah-wi′e.
Hayti, hay′te.
Haytien, hay′te-en.
Heidelberg, hi′del-berg.
Herat, her-aht′.
Hiawatha, hi-ah-wah′thah.
Himalay′a.
Hindoostan′.
He-ang′ Ho.
Ho′bo-ken.
Holstein, hol′stine.
Hondo, hone′do.
Honolu′lu.
Housatonic, hoo-sa-ton′ik.
Hue, hoo-ay′.
Hu′ron.
Hy-der-a-bad′.

I.

I′daho.
Ilamani, eel-yah-mah′ne.
Illinois, il-lin-oy′.
Indies, in′diz.
Indighirca, in-de-ghir′kah.
Inn-pruck, ins′pruck.
Inverness′.
I′owa.
Irkoutsk′.
Iroquois, ir-o-kwoy′.
Ir′tysh.
Isar, ee′zer.
Ispahan′.
Iviça, e-ve′sah.

J.

Jalapa, hah-lah′pah.
Jeddo, yed′do.
Jesso, yes′so, Properly Yezo.
Jorullo, ho-rool′yo.
Jungfrau, yoong′frow.
Juniata, joo-ne-at′tah.

K.

Kal-a-ma-zoo′.
Kamtschatka, kam-chat′ka.
Kanagawa, kan-a-gaw′wah.
Kanawha, ka-naw′wa.
Kankakee′.
Kara, kah′rah.
Ka-tah′din.
Kehl, kail.
Kelat′.
Ken-ne-bec′.
Ke′o-kuk.
Khartoum′.
Khiva, kee′vah.
Khokan, ko-kahn′.
Kiang-ku, ke-ang-kew′.
Kickapoo′.
Kiel, keel.
Kiushiu, ke-oo′she-oo′.
Koko-Nor′.
Kordofan, kor-do-fahn′.
Kuen-Lun, kwen-loon′.

L.

Lab-ra-dor′.
Lachine, lah-sheen′.
Ladakh, lah-dakh.
Lago Maggiore, lah′go mahd-jo′ra.
La Guayra, lah gwi′rah.
Lahore′.
Lanc′aster.
Languedoc, lon-geh-dok′.
Laos, lah′ose.
La Plata, lah-plah′tah.
La Porte, lah-port′.
La Rue, lah-roo′.
La Salle, lah-sal′.
Lech, lay.
Leicester, les′ter.
Leigh, lee.
Leipsic, lipe′sik.
Leyden, li′den.
Lichtenfels, lik′ten-fels.
Lille, or Lisle, leel.
Lima (Peru), le′mah; (U. S.), li′ma.
Limoges, li′ne.
Lipari, lip′a-re.
Llanos, l′yah′noce.
Loch Lomond, lok-lo′mund.
Loffo′den.
Loire, lwahr.
Liu Kiu.
Lough Neagh, loh-nay′.
Louis, St., sent loo′is, or loo′e.
Louisiade, loo-e-ze-ahd′.
Louisville, loo′is-vil.
Lowell, lo′el.
Lucayos, loo-ki′os.
Lupata, lu-pah′tah.
Luzon, loo-zone′.
Lyonnais, le-on-nay′.

M.

Macao, mah-kah′o, or mah-kow′.
Mack′inaw.
Madeira, ma-de′rah.
Madras′.

Madrid' (U.S.); mah-dreed' (Sp.)
Mael'strom, mayl'strum.
Mageroe, mag'er-o.
Maggiore, mahd-jo'ra.
Mal'aga, or mah'lah-gah.
Ma-lay'.
Mal'dives
Managua, mah-nah'gwah.
Manitouwoc, man-e-too-wok'.
Man'tu-a.
Manzanillo, mahn-zah-neel'yo.
Mapocho, mah-po'cho.
Maracaybo, mah-rah-ki'bo.
Maranham'.
Maranon, mah-rahn-yon'.
Mar'mo-ra.
Marquesas, mar-kay'sahs.
Marseilles, mahr-sails'.
Marguerite, mar-geh-reet'.
Martinique, mar-ti-neek'.
Ma-tan'zas.
Matamee', mata-mi'.
Mauch Chunk, mawk-chunk'.
Mauritius, maw-rish'e-us.
Maysi, mi'se.
Mazatlan'.
Medina, me-de'nah.
Meiga, megz.
Moinam, me-e-nam'.
Melbourne, mel'burn.
Menai, men'i, or men'ay.
Me-nan'.
Mendocino, men-do-se'no.
Menominee.
Mequinez, mek'i-nez.
Merida, mer'e-dah
Merthyr Tydvil, mer'ther tid'vil.
Messina, mess-se'nah.
Meuse, muze.
Mezene, mez-ane'.
Miako, me-ah'ko.
Mich'i-gan.
Michilimackinac, mich-il-e-mak'e-naw
Mil'an, or mi-lahn'.
Mille Lacs, meel-lahk'.
Minatit'lan.
Mincio, min'tcho.
Mindanao, min-dah-nah'o.
Minnehaha, min-ne-hah'hah.
Miramichee, mir-a-ma-shee'.
Mobile, mo-beel'.
Mocha, mo'kah.
Mod'e-na.
Mog-a-dore'.
Molino-del-Rey, mo-le'no-del-ray.
Mom-pox, mom-posh'.
Monaco, mo-nah'ko.
Mon-cho-bo'.
Montana, mon-tah'nah.
Mont Blanc, mon-blon.
Mont Cenis, seh-ne'.
Montenegro, mon-ta-nay'gro.
Monterey, mon-ta-ray'.
Monte Video, mon-ta-ve'da-o.
Montserrat'.
Moscow, mos'ko.
Moselle, mo-zel'.
Mourzouk, moor-zook'.
Munich, mu'nik.
Musent'.
Muscatine, mus-ka-teen'.
Muscogee, g hard.
Muske'gon.

N.

Nahant'.
Nagasaki, nahg-ga-sah'ke.
Nankin'.
Napo, nah'po.
Nash'ua.
Nauvoo'.
Navarino, nah-vah-re'no.
Neagh, nay.
Negropont, nay'gro-pont.
Nep'is-sing.
Neufchatel, nush-ah-tel'.
Nevada, nay-vah'dah.
New'foundland'.
New Or'leans.
Ngami, n'gah'me.
Nice, neece.
Nic'o-bar.
Niger, ni'jer.
Nippon, nip'on.
Nori, nor.
Nottingham, not'ting-am.
Norfolk, nor'fok.
Norwich (Eng.), nor'ij; (U.S.), nor-rich.
Novgorod'.
Nueces, nway'ses.
Nuevitas, nway-ve'tas.
Nyassi, nyahs'se.

O.

Oahu, wah'hoo.
O'a-sis.
Oaxaca, wah-hah'kah.
Obi, o'be.
Obidos, o-be'dos.
Ocmul'gee, g hard.
Oco'nee.
O'cracoke.
Ogee'chee.
O-kee-cho'bee.
O-ke-fin-o'ke.
O-kee-cho'bee.
Okhotsk, o-kotsk'.
Omaha, o'ma-haw.
Oneida, o-ni'da.
Onondaga, on-un-daw'ga.
Ontonagon, on-ton-ah'gon.
Opelousas, op-e-loo'sas.
Orizaba, o-re-sah'bah.
Or'le-ans.
Ortegal, or-te-gahl'.
Osh'kosh.
Otaheite, o-tah-he'te.
Oude, ood.
Ouse, ooz.
Oviedo, o-ve-ay'do.
Owyhee, o-wi'e.
O-zark'.

P.

Padua, pad'yu-ah.
Pais'ley.
Pal'es-tine.
Palo Alto, pah'lo ahl'to.
Palos, pah'loce.
Panama, pah-nah-mah'.
Paoli, pa-o'le.
Papua, pap'oo-a.
Para, pah-rah'.
Parana, pah-rah-nah'.
Paria, pah're-ah.
Parima, pah-re'mah.
Paso, pah'so.
Patos, pah'toce.
Paz, La, lah-pahz.
Pe-che-lee'.
Pecos, pay'koos.
Peeks'kill.
Pegu, pe-goo'.
Pelho, pay'ho.
Pelpus, pay'e-pus.
Pekin'.
Pelee, peh-lay'.
Pelew, pe-loo'.
Pelung, pay-lung'.
Pem'bina.
Pepin, pip'in.
Perdido, per-de'do.
Pernambuco, per-nahm-boo'co.
Perouse, pe-rooz'.
Peru, pe-roo'.
Pesth, pest.
Petcholee'.
Petch'ora.
Philippine, fil'ip-pin.
Pichincha, pe-cheen'chah.
Pictou, pik-too'.
Pied'mont.
Pierre, pe-air'.
Pilcomayo, pil-ko-mi'o.
Pinos, pe'nos.
Pisa, pe'sah.
Pisgah, piz'gah.
Plata, La, plah'tah.
Plateau du Coteau, plah-to' du ko-to'.
Pocahontas.
Point Coupee, koo-pee'.
Pollinao, po-leen-yahk'.
Pompeii, pom-pay'e.
Pondicherry, pon-de-sher're.
Pontchartrain, pon-shar-train'.
Popayan, po-pi-ahn'.
Po-po-cat-a-petl'.
Porto Praya, pri'ah.
Porto Rico, re'ko.
Port Paix, por-pay'.
Potosi, po-to-se'.
Poughkeepsie, po-kip'se.
Powhatan'.
Prairie du Chien, du sheen.
Presque Isle, presk-eel'.
Puebla, pweb'lah.
Puerto Principe, pwer'to-preen'se-pay.
Puget, pu'jet.
Pulaski, pu-las'ke.
Punjaub'.
Putumayo, poo-too-mi'o.
Pyr'enees.

Q.

Quebec'.
Quillota, keel-yo'tah.
Quiloa, kee'lo-a.
Quin-e-baug'.
Quito, kee'to.
Quogue, kwōg.

R.

Racine, Ras-seen'.
Rahway, raw'way.
Raleigh, raw'le.
Rangoon, rahn-goon'.
Rapidan'.
Raven'na.
Reading, red'ing.
Reggio, red'jo.
Reikiavik, ri'ke-a-vik.
Reims, or Rheims, reems.
Rensselaer, ren'se-ler.
Resaca de la Palma, re-sah'kah da lah pal'mah.
Restigouche, res-to-goo-shay'.
Reus, ray'oose.
Revillagigedo, ray-veel-yah-he-hay'do
Reyes, ray'es.
Rhaetian, re'shun.
Rhea, ray.
Richelieu, re-she-loo'.
Rideau, re-do'.
Riesengebirge, re'sen-ge-beer'geh.
Ri'ga, or re'ga.
Rio Dulce, re'o-dool'sa.
Rio Grande, re'o-grahn'day.
Rio Janeiro, re'o-ja-ne'ro.
Ristigouche, ris-to-goo'sha.
Rochelle, ro-shell'.
Romagna, ro-mahn'ya.
Rouen, roo'en.
Russia, rush'e-a, or roo'she-a.
Ryswick, riz'wik.

S.

Saar'dam.
Sabine, sah-been'.
Sac, sawk.
Sackatoo, sahk-kah-too'.
Saco, saw'ko.
Saghalien, sah-gah'le-en.
Sagitano, sahl-yah'no.
Sagua-la-Grande, sah'gwah-lah-grahn'day.
Saguenay, sag-eh-nay'.
Sahara, sa-hah'rah.
Saigon, si-gon'.
Salado, sah-lah'do.
Salisbury, sawlz'ber-e.
Salonica, sah-lo-ne'kah.
Saltillo, sahl-teel'yo.
Salvador'.
Samana, sah-mah-nah'.
Samarcand'.
Sana, sah-nah'.
San Bernardino, sahn ber-nar-de'no.
San Blas, san blahs'.
San Diego, san de-ay'go.
San Felipe, fa-le'pay.
Sangamon, Sang'ga-mon.
San Joaquin, wah-keen'.
San Jose, ho-say'.
San Juan, hoo-ahn'.
San Sal-va-dor'.
Santa Cruz, sahn'ta-krooz.
Santa Fe, fay.
Santa Margarita, mahr-gah-ree'tah.
Santiago, sahn-te-ah'go.
Saone, sone.
Saratof'.
Sas-katch'e-wan.
Saugatuck.'
Saugerties, saw'ger-tees.
Scheidt, skelt.
Schiedam, ske-daum'.
Schodak'.
Schoo'dik.
Schuyler, ski'ler.
Schuylkill, skool'kill.
Schwerin, shway-reen'.
Scinde.
Scio, si'o.
Sciota, si-o'to.
Scutari, skoo-tah'ree.
Se-ba'go.
Sego, say'go.
Seidlitz, side'litz.
Seine, sane.
Senegal, sen-e-gawl'.
Serapis, ser-ay'pis.
Sergipe del Rey, ser-she'pay-del-ray.
Sevier, se-veer'.
Seychelles, sa-shell'.
Shamo, sha-mo'.
Shang-Hae, shang-hi'.
Shawangunk, shong'gum.
She-boy'gan.
Shen-an-do'ah.
Shiraz, she-rahz'.
Shosh-ones'.
Sierra Leone, se-er'rah le-o'na.
Sierra Madre, mah'dray.
Sierra Nevada, nay-vah'dah.
Singapore, sing-gah-pore'.
Sigourney, sig'ur-ne.
Si-koke'.
Sinai, si'nay.
Sin-o'pe.
Sioot, or Siout, se-oot'.
Sioux, soo.
Sir-i-Kul, seer-e-kol'.
Sisal, se-sahl'.
Skan-e-at'e-les.
Skye, ski.
Sli'go.
So-co'tra, or sok'o-trah.
Solferino, sol-fa-re'no.
Sombrero, som-bra'ro.
Soodan', or Soudan'.
Soo-loo'.
Sorato, so-rah'tah.
So-rel'.
Soutcheou, soo-che-oo'.
Spa, spah.
Spree, spray.
Stamboul'.
St. Ber-nard'.
Stettin, stet-teen'.
Stromboli, strom'bo-le.
Stutt'gart.
Su'ez.
Suffolk, suf'fok.
Sumatra, soo-mah'trah.
Sumbawa, soo-baw'wah.
Surinam, soo-rin-am'.
Suwanee, soo-wah'ne.
Swansea, swon'se.
Syr'a-cuse.

T.

Tabreez, or Tabriz, tah-breez'.
Ta-co'ma.
Tahiti, tah-he'te.
Tah'le-quah.
Taliaferro, tol'e-ver.
Tamaqua, ta-maw'kwa.
Tamaulipas, tah-mou-le'pahs.
Tampico, tahm-pe'ko.
Tananarivoo, tah-nah-nah-re-voo'.
Taney, taw'ne.
Tanganyika, tahn-gan-ye'kah.
Tangier, tan-jeer'.
Taos, tah'ose.
Tapajos, tah-pah'-zhoce.
Tarifa, ta-re'fah.
Taunton (Mass.), tahn'ton.
Tchad, chahd.
Teheran, teh-her-ahn'.
Tehuantepec, tay-won-tay-pek'.
Terre Bon, tare-bon'.
Terre Haute', ter-reh-hote'.
Teulada, te-oo-lah'dah.
Tescuco, tes-koo'ko.
Theiss, tale'bo.
Thames, temz.
Theiss, tise.
Thes-sa-lo-ni'ca.
Thibet, tib'et.
Ticino, te-chee'no.
Tientsin, te-ent'seen.
Tim-boo'too.
Timor, te-more'.
Titicaca, tit-e-kah'kah.
Tivoli, tiv'o-le.
Tokio, to'ke-o.
Toaquin, too-keen'.
To-pe'ka.
Toulon'.
Toulouse, too-looz'.
Traf-al-gar'.
Treb'i-zond.
Tref'a-gar.
Tre-mont'.
Trieste, tre-est'.
Trin-i-dad'.
Tripoli, trip'o-le.
Truxillo, troo'heel-yo.
Tu-at'.
Tuaricks, too-ah-reeks'.
Tubac, too-bahk'.
Tulare, too-lah'ray.
Tungpusas, toong-goo'sez.
Tunguragua, toong-goo-rah'gwa.
Tu'rin, or tu-rin'.

U.

Ucayale, oo-ki-ah'lay.
Ujija, oo-je'ja.
Unadilla, yoo-nah-dil'lah.
Upernavik, oo-per'nah-vik.
Utrecht, u'trekt.

V.

Valais, vah-lay'.
Valdai, vahl'di.
Val'la-do-lid'.
Vallejo, vahl-ya'ho.
Valparaiso, vahl-pah-ri'so.
Van Diemen's, de'manz.
Vera Cruz, vay'rah krooz.
Vergennes, ver-jenz'.
Vermejo, ver-may'ho.
Versailles, ver-salle'.
Ve-vay'.
Vienna, ve-en'na.
Villa Real, veel'yah re-ahl'.
Vincennes, vin-senz'.
Vindhya, vind'yah.
Vist'u-la.
Viviers, ve-ve-ay'.

W.

Wachusett, waw-choo'set.
Wallachia, wol-lay'ki-a.
Wartha, wahr'tah.
Washita, wash'e-taw.
Wieliczka, we-litch'ka.
Wilkesbarre, wilks'bar-re.
Willa'mette.
Winneba'go.
Winnipiseogee, win-e-pe-saw'ke.
Winona, we-no'nah.
Worcester, woos'ter.
Wy-o'ming.

Y.

Yakoutsk, yah-kootsk'.
Yang-tse-kiang, ke-ang'.
Yank'ton.
Yap-hank.
Ya-zoo'.
Yenikale, yen-e-kah'lay.
Yenisei, yen-e-say'e.
Yeniseisk, yen-e-say'isk.
Youghiogheny, yoh-ho-gay'ne.
Ypsilanti, ip-si-lan'te.

Z.

Zacatecas, zah-kah-tay-kas.
Zam-beze', or zam-bay'ze.
Zanguibar, zang-ge-bar'.
Zanzibar'.
Zurich, zoo'rik.
Zuyder Zee, zi'der-zee.

MEANING OF GEOGRAPHICAL NAMES.

Abyssinia, a mixed people.
Aix-la-Chapelle, waters of the chapel.
Alps, snow-clad mountains.
Antigua, ancient.
Arizona, sand hills.
Asia, the east.
Azores, hawks.
Bab-el-Mandeb, gate of tears.
Bahia Honda, deep bay.
Baton Rouge, red staff.
Bayou, a creek.
Belleisle, beautiful island.
Ben Lomond, beacon mountain.
Ben More, great mountain.
Blanc or Blanco, white.
Bombay, good harbor.
Bon Homme, good man.
Bordeaux, border of the water.
Bras d'Or, an arm of gold.
Buena Vista, fine view.
Buenos Ayres, fine air.
Cairo, victorious.
Cape Verd, green cape.
Catskill, cat's or lynx creek.
Caucasus, white mountains.
Cayuga, long pond.
Cerro Gordo, mountain pass.
Charleston, after Charles I., of England.
Chesapeake, great waters.
Chili, land of snow.
Chimborazo, chimney.
Chuquisaca, golden bridge.
Clermont, clear mountain.
Colorado, red or colored.
Cork, marsh.
Costa Rica, rich coast.
Cumberland, a land of hollows.
Delaware, after Lord de la Ware.
Des Moines, a place of mounds.
Dnieper, upper river.
Dniester, lower river.
Dwina, double river.
Ebro, foaming river.
Elbe, white.
El Paso, the pass.
Espiritu Santo, Holy Spirit.
Ethiopia, to burn.
Finisterre, end of the land.
Florence, flowery city.
Fond du Lac, end of the lake.
Frio, cold.
Fuego, fire.
Galapagos, tortoises.
Glasgow, dark ravine.
Gracias à Dios, thanks to God.
Havre de Grace, harbor of safety.
Hayti, high land.
Henlopen, run in.
Hudson, after Henry Hudson.
Irrawaddy, great river.
Java, rice.
Jerusalem, place of peace.
Jordan, the flowing.
Kansas, smoky water.
Katahdin, highest place.
Kennebec, long lake.
Liberia, free.
Louisiana, after Louis XIV. of France.
Majorca, greater.
Manitoulin, spirit islands.
Mediterranean, middle of the land.
Minnehaha, laughing water.
Mississippi, great river.
Missouri, muddy.
Minorca, less.
Montreal, royal mountain.
Nova Scotia, new Scotland.
Palestine, land of wanderers.
Patagonia, clumsy feet.
Piedmont, foot of the mountain.
Polynesia, many islands.
San Domingo, holy Sabbath.
San Salvador, holy Saviour.
Santa Cruz or Croix, holy cross.
Santa Fe, holy faith.

ILLINOIS, IOWA, AND MISSOURI.

MAP-DRAWING.

(To precede the Questions on page 42.)

Draw the eastern boundary line of ***Illinois,*** the same as the western boundary line of ***Indiana,*** ¾ m. from Lake Michigan to Wabash River. From F, measure ¼ m. north and ⅛ m. west to A, and draw the southern coast of ***Lake Michigan.*** Locate ***Chicago.***

Mark the northern boundary ¾ m. from A to B. Locate ***Cairo*** 1⅞ m. south of the line A B. Measure the extreme breadth of the State 1¼ m. on the line V U, and draw the ***Mississippi, Ohio,*** and ***Wabash Rivers.*** Complete drawing of State.

Next draw ***Iowa,*** commencing with its northern boundary 1⅝ m. from C to D, and ⅛ m. north of the northern boundary of ***Illinois.*** Observe the measurements, and complete the drawing of Iowa.

Complete ***Missouri,*** by commencing at K and measuring ¼ m. east and ½ m. south to ***Kansas City.*** Mark its southern boundary 1⅜ m. south of its northern, 1½ m. in length from N to T, and ¼ m. from T to S.

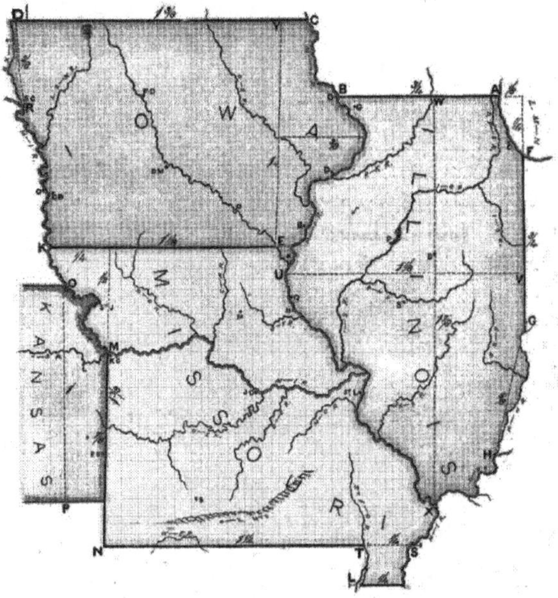

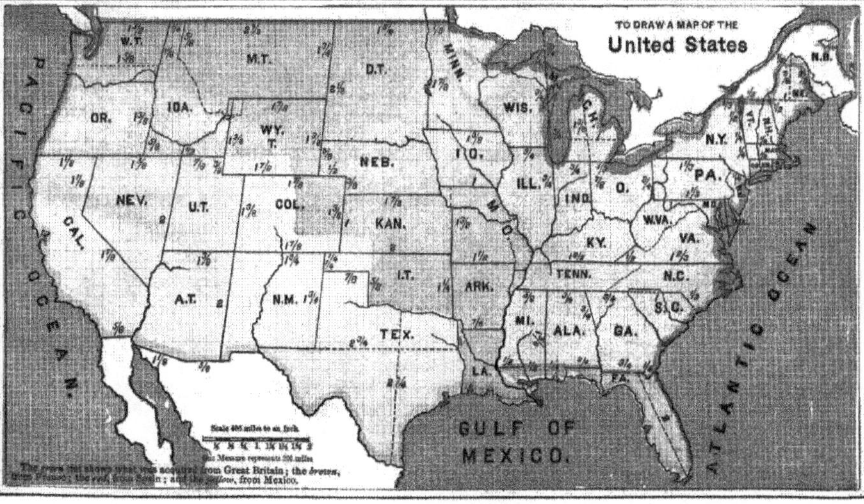

HEIGHTS OF MOUNTAINS.

ASIA.

Mountains.	Situation.	Elevation in feet.
Mt. Everest	in the Himalaya Chain, the highest point on the globe	29,000
Mt. Kunchinginga	in the Himalaya Chain	28,178
Mt. Dhawala-giri	" " "	28,000
Mt. Choumalarte	" " "	23,929
Mt. Hindoo Koh	Hindoo Coosh Mts.	20,230
Thian-Shan Mts.	Chinese Empire	20,000
Kuen-Lun Mts.	North of Thibet ... 15,000 to	18,000
Mt. Ararat		17,210
Mt. Kliutshewskaja	Kamtschatka	15,763
Mt. Damavend	Elburz Mts., south of the Caspian Sea	14,700
Mt. Arjish	Anti-Taurus Mts., Asia Minor	13,000
Soliman Mts.	between Hindostan and Afghanistan	12,000
Lebanon Mts.	Syria	12,000
Altai Mts.	highest	10,300
Taurus Mts.	highest	10,000
Mt. Hermon	highest of Anti-Lebanon Mts., Syria	10,000
Mt. Sinai	Arabia	7,497

SOUTH AMERICA.

Vol. Aconcagua	east of Chili	23,167
Mt. Sehama	south part of Peru	22,350
Vol. Gualatieri	" " "	22,000
Mt. Chimborazo	Ecuador	21,424
Mt. Sorata	Bolivia	21,286
Mt. Illimani	"	21,149
Mt. Cuquihamba	Peru	21,000
Vol. Arequipa	"	20,320
Vol. Atacama	Bolivia	18,000
Vol. Antisana	Ecuador	19,132
Vol. Cotopaxi	"	18,887
Vol. Tolima	Colombia	18,000
Vol. Pichincha	Ecuador	15,924
Andes of Chili	highest	16,000
Andes of Patagonia	"	8,000
Cape Horn		500
Height of Snow Line	in Colombia	15,800
" " "	in Bolivia ... 16,000 to	18,000
" " "	in Chili	8,000
" " "	in Patagonia	3,000

NORTH AMERICA.

Vol. Popocatepetl	Mexico	18,500
Mt. St. Elias	Alaska ... about	18,000
Vol. Orizaba	Mexico	17,371
Mt. Brown	British America	15,900
Mt. Hooker	" "	15,675
Mt. Whitney	California	15,086
Mt. Fairweather	Alaska	14,900
Pike's Peak	Colorado	14,500
Mt. Shasta	California	14,442
Mt. Tyndall	"	14,386
Mt. Kaweah	"	14,000
Fremont's Peak	Wyoming Territory	13,570
Mt. St. Helens	Washington Territory	13,300
Mt. Dana	California	13,227
Mt. Lyell	"	13,217
Long's Peak	Colorado	12,505
Mt. Hood	Oregon	11,000
Mt. Linn	California	10,000
Mt. San Bernardino	"	8,570
Mt. Baker	Washington Territory	
Mt. Ranier	" "	
Mt. Adams	" "	
Laramie Peak	Wyoming Territory	8,000
Blue Mts.	Jamaica	7,300
Sierra del Cobre	Cuba—highest	7,200
Mitchell's Peak	North Carolina	6,707
Mt. Washington	New Hampshire	6,428
Oraefa Jokull	Iceland	6,409
Mt. Bailey	California	6,357
Mt. Pierce	"	6,000
Vol. Hecla	Iceland	5,700
Vol. Soufriere	Guadaloupe	5,550
Mt. Marcy	New York	5,379
Mt. Katahdin	Maine	5,300
Mt. Hamilton	California	4,450
Mt. Mansfield	Vermont	4,430
Peaks of Otter	Virginia	4,260
Camel's Hump	Vermont	4,188
Saddleback Mt.	Maine	4,000
Mt. Diablo	California	3,876
Round Top	New York	3,804
Alleghany Mts.	average	2,580
Wachusett Mt.	Massachusetts	2,030
Highlands	New York	1,600
Pilot Knob	Missouri	1,500
Mt. Tom	Massachusetts	1,200

EUROPE.

Mountains.	Situation.	Elevation in feet
Mt. Elboorz	highest of Caucasus	17,796
Mt. Blanc	" " Alps	15,810
Mt. Rosa	" in Switzerland	15,208
Mt. Cervin	between Switzerland and Italy	14,771
Mt. Pelvoux	France	14,108
Finster Aarhorn	Switzerland	14,026
Mt. Viso	between France and Italy	13,599
Ortler Spitz	Austria	12,811
Mt. Mulahacen	Spain	11,660
Mt. Genevre		11,614
The Simplon	between Switzerland and Italy	11,541
Mt. Cenis	France	11,460
Maladetta	highest of Pyrenees	11,168
Peak of Valetta		11,389
Great St. Bernard	Switzerland	11,080
Cantabrian Mts.	Spain—highest	10,998
Mt. Perdu	Pyrenees	10,994
Vol. Etna	Sicily	10,874
Mt. Cuccio	"	10,871
Mt. St. Gothard	Switzerland	10,595
Castile Mts.	Spain	10,551
Mt. Olympus	Turkey	9,745
Mt. Korn	"	9,678
Mt. Athos	Greece	9,628
Pic du Midi	Pyrenees	9,540
Pindus Mts.		8,950
Mt. Lommitz	Austria	8,779
Monte Rotondo	Corsica	8,763
Mt. Guione	Greece	8,239
Mt. Parnassus	"	8,068
Scandinavian Mts.	highest	8,152
Pass of Venasque	Pyrenees	7,917
Rhodope Mts.	Turkey	7,800
Sierra d'Estrella	Portugal	7,524
Mt. Gennargentu	Sardinia	7,000
Jura Mts.	France and Sicily	6,588
Mt. D'Or	France	6,188
Cevennes Mts.	"	5,794
Sierra Morena	Spain	5,560
Mt. Obdorsk		5,397
Fichtel Gebirge	Germany	5,356
Riesen Gebirge	"	5,280
Ural Mts.	highest	5,275
Vosges	France	4,686
Black Forest	Baden	4,675
Bohemian Mts.	Austria	4,613
Erz Gebirge	"	4,560
Ben Nevis	Scotland—highest in Great Britain	4,406
Ben Mac Dhu	"	4,327
Cairngorm	"	4,095
Ben Avlow	"	4,000
Vol. Vesuvius	Italy	3,948
Ben Lawers	Scotland	3,945
Brocken Mt.	Hartz Mts., Germany	3,740
Ben Wyvis	Scotland	3,720
Snowdon Mt.	Wales	3,571
Carn Tual	highest in Ireland	3,414
Sca Fell	England	3,092
Helvellyn	"	3,055
Skiddaw	"	3,022
Moravian Mts.	Austria	3,006
Height of Snow Line	Sierra Nevada, Spain	11,200
" " "	Caucasus Mts.	11,000
" " "	Sicily	9,500
" " "	Mt. Olympus	9,000
" " "	Swiss Alps	8,900
" " "	Central Italy	8,400
" " "	Pyrenees	8,000
" " "	Carpathians	6,000
" " "	Scandinavian Mts.	5,000
" " "	Grampians	4,500
" " "	North Cape	2,400

AFRICA.

Mt. Kenia	Eastern Africa	20,000
Mt. Kilimandjaro	" "	20,000
Abba Jared	Abyssinia	15,000
Cameroon Mts.	Lower Guinea	13,000
Peak of Teneriffe, Vol.	Canary Islands	12,182
Mt. Miltsin	Morocco	11,400
Red Mts.	Madagascar	11,000
Mt. Spitzkop	Cape Colony	10,250
Table Mt.	Cape of Good Hope	3,672

OCEANICA.

Mt. Ophir	Sumatra	13,842
Mt. Kini Balu	Borneo	13,698
Vol. Mauna Loa	Sandwich Islands	13,120
Mt. Semero	Java	12,000
Mt. Erebus	Victoria Land	12,400
Mt. Kosciusko	Australia	6,500
Mt. Humboldt	Tasmania	5,520

HEIGHTS OF SOME INHABITED SITES.

Names.	Situation.	Feet above sea level.
Runechuasi	Peru	15,640
Tacora	Village in Peru	13,690
Antisana	Shepherds' huts, Ecuador	13,454
Potosi	City in Bolivia	13,330
Puno	City in Peru	12,870
Cuzco	Bolivia	11,380
Leh	City in Thibet	[illegible]
Quito	Ecuador	9,543
Sherman	highest point on Pacific R. R.	8,242
Hospice of Gt. St. Bernard	Alps	7,963
Arequipa	City in Peru	7,552
Mexico	City	7,400
Cabul	Afghanistan	6,396
Ispahan	Persia	4,140
Jerusalem	Palestine	2,500
Madrid	Spain	1,995
Munich	Bavaria	1,754
Geneva	Switzerland	1,230

DISTANCES AT WHICH MOUNTAINS HAVE BEEN SEEN.

Mountains.	Miles.
Himalaya Mts.	244
Mt. Ararat	240
Mt. Chimborazo	163
Peak of Teneriffe	135
Mt. Athos, Greece	100

SETTLEMENT OF STATES.

	Settled.	Admitted.
New Mexico	1540	
Florida	1565	March 3, 1845
Virginia	1607	June 26, 1788
New York	1614	July 26, 1788
Massachusetts	1620	Feb. 6, 1788
New Jersey	1630	Dec. 18, 1787
New Hampshire	1624	June 21, 1788
Maine	1630	March 15, 1820
Connecticut	1633	Jan. 9, 1788
Maryland	1634	April 28, 1788
Rhode Island	1636	May 29, 1790
Delaware	1638	Dec. 7, 1787
Pennsylvania	1643	Dec. 12, 1787
North Carolina	1650	Nov. 21, 1789
Wisconsin	1669	May 29, 1848
South Carolina	1670	May 23, 1788
Michigan	1670	Jan. 26, 1837
Illinois	1683	Dec. 3, 1818
Arkansas	1685	June 15, 1836
Indiana	1690	Dec. 11, 1816
Texas	1692	Dec. 29, 1845
Louisiana	1699	April 8, 1812
Alabama	1711	Dec. 14, 1819
Mississippi	1716	Dec. 10, 1817
Vermont	1725	March 4, 1791
Georgia	1733	Jan. 2, 1788
Missouri	1755	Aug. 10, 1821
Tennessee	1757	June 1, 1796
California	1768	Sept. 9, 1850
Kentucky	1770	June 1, 1792
Ohio	1788	Nov. 29, 1802
Oregon	1811	Aug. 14, 1848
Washington	1811	Aug. 14, 1848
Iowa	1833	
Minnesota	1846	March 3, 1849
Utah	1848	
Nebraska		May 30, 1854
Kansas		May 30, 1854
Colorado		
Montana		

RIVERS OF THE WORLD.

NORTH AMERICA.

Names.	Length in m.	Names.	Length in m.
Missouri, to the Mississippi	3,100	Colorado (of the West)	1,000
Missouri, to the G. of Mexico	4,350	Ohio	948
Mississippi	3,160	Lewis, or Snake	900
Mackenzie's, fr. Slave Lake	900	Tennessee	800
Mackenzie's, fr. head of the Athabasca	2,440	Cumberland	600
St. Lawrence, from Lake Ontario	750	Appalachicola	600
St. Lawrence, from head of St. Louis R.	2,200	Mobile	550
Arkansas	2,170	Susquehannah	450
Rio Grande	1,800	James	450
Red	1,500	Sacramento	420
Platte, or Nebraska	1,500	Potomac	400
Nelson and Saskatchewan	1,600	Savannah	400
Columbia, or Oregon	1,200	Altamaha	400
		Connecticut	400
		Minnesota	375
		Roanoke	350
		Pedee	350
		Hudson	325
		Delaware	300
		Penobscot	300
		Mohawk	160
		Kennebec	150
		Genesee	145
		Merrimac	110

SOUTH AMERICA.

Amazon	4,000	Orinoco	1,550
La Plata, from the head of the Parana	2,250	St. Francisco	1,200
Madeira	1,800	Tocantins	1,100
Paraguay	1,800	Araguay	1,000
		Purus	900
		Magdalena	800

EUROPE.

Volga	2,000
Danube	1,725
Dnieper	1,230
Don	985
Rhine	900
Petchora	900
Ural	840
Tagus	650
Rhone	648
Loire	580

ASIA.

Yang-tse-Kiang	2,500
Lena	2,400
Yenisei	2,300
Amoor	2,200
Obi	2,000
Hoang Ho	2,000
Cambodia	2,000
Irtysh	1,700

AFRICA.

Nile	4,000
Niger	2,700
Zambese	1,350

TENNESSEE,

Places represented on the map by numbers.

1 Troy.
2 Dyersburg.
3 Paris.
4 Huntingdon.
5 Dover.
6 Waverly.
7 Ashland.
8 Smithville.
9 Livingston.
10 Jamestown.
11 Huntsville.
12 Maynardsville.
13 Tazewell.
14 Rogersville.
15 Taylorsville.
16 Elizabethtown.
17 Kingston.
18 Madisonville.
19 Athens.
20 Decatur.
21 Dunlap.
22 McMinnville.
23 Manchester.
24 Waynesboro.
25 Decaturville.
26 Savannah.
27 Purdy.
28 Bolivar.

NAMES OF CITIES AND TOWNS REPRESENTED ON THE MAPS BY NUMBERS.

ALABAMA.
1 Tuscumbia.
2 Russellville.
3 Moulton.
4 Bellefonte.
5 Lebanon.
6 Blountsville.
7 Center.
8 Jacksonville.
9 Ashville.
10 Jasper.
11 Pikeville.
12 Fayetteville.
13 Ashland.
14 Wedowee.
15 Rockford.
16 Marion.
17 Linden.
18 Crawford.
19 Butler.
20 Grove Hill.
21 St. Stephens.
22 Monroeville.
23 Troy.
24 Rutledge.
25 Ozark.
26 Elba.
27 Newton.
28 Andalusia.
29 Sparta.

ARKANSAS.
1 Bentonville.
2 Carrollton.
3 Jasper.
4 Lebanon.
5 Sherman.
6 Pilot Hill.
7 Smithville.
8 Mt. Olive.
9 Gainesville.
10 Osceola.
11 Jonesboro
12 Harrisburg.
13 Batesville.
14 Jacksonport.
15 Clinton.
16 Ozark.
17 Clarksville.
18 Dover.
19 Lewisburg.
20 Augusta.
21 Marion.
22 Mt. Vernon.
23 Clarendon.
24 De Witt.
25 Brownsville.
26 Perryville.
27 Danville.
28 Mt. Ida.
29 Dallas.
30 Benton.
31 Rockport.
32 Princeton.
33 Arkansas Post
34 Napoleon.
35 Monticello.
36 Lake Village.
37 Hamburg.
38 Hampton.
39 El Dorado.
40 Calhoun.
41 Lewisville.
42 Washington.
43 Lockesburg.

CONN.
1 Suffield.
2 Vernon.
3 Colchester.
4 Windham.
5 Putnam.
6 Woodstock.
7 Derby.
8 Branford.
9 Guilford.

CALIFORNIA.
1 Santa Cruz.
2 Eureka.
3 Napa City.
4 Gilroy.
5 Brooklyn.
6 Placerville.
7 Oroville.
8 Sonora.
9 Watsonville.
10 Columbia.
11 Monterey.
12 San Pablo.
13 Yreka City.
14 Colusa.
15 Orleans Bar.
16 Weaverville.
17 Shasta City.
18 Susanville.
19 Quincy.
20 Red Bluff.
21 Downieville.
22 Lakeport.
23 Mariposa.
24 Snelling.
25 Millerton.
26 Keysville.
27 Havilah.

DAKOTA.
1 Medary.
2 Maxwell.
3 Sivan Lake.
4 Sioux Falls.
5 Bon Homme.
6 Vermillion.
7 Elk River.

FLORIDA.
1 Uchee Anna.
2 Cerro Gordo.
3 Holmes Valley.
4 Marianna.
5 Newport.
6 Newnansville.
7 Wacassassa.
8 Adamsville.
9 Bayport.
10 Enterprise.
11 Tampa.
12 Manatee.
13 St. Lucie.
14 Key Biscayne.

GEORGIA.
1 Trenton.
2 Ringgold.
3 La Fayette.
4 Summerville.
5 Morganton.
6 Blairsville.
7 Dahlonega.
8 Clarksville.
9 Carnesville.
10 Gainesville.
11 Lawrenceville.
12 Jefferson.
13 Elberton.
14 Lincolnton.
15 Lexington.
16 Washington.
17 Watkinsville.
18 Monroe.
19 Tallapoosa.
20 Campbellton
21 Carrollton.
22 Franklin.
23 McDonough.
24 Covington
25 Greensboro.
26 Warrenton.
27 Eatonton.
28 Jackson.
29 Thomaston.
30 Hamilton.
31 Talbotton.
32 Sandersville.
33 Louisville.
34 Waynesboro.
35 Sylvania.
36 Statesboro.
37 Reidsville.
38 Mt. Vernon.
39 Hawkinsville.
40 Jacksonville
41 Holmesville.
42 Irwinville.
43 Brunswick.
44 Waresboro.
45 Magnolia.
46 Nashville.
47 Troupville.
48 Moultrie.
49 Albany.
50 Camilla.
51 Bainbridge.
52 Colquit.
53 Blakely.
54 Cedartown.

ILLINOIS.
1 Belvidere.
2 Woodstock.
3 Mt. Carroll.
4 Oregon.
5 Sycamore.
6 Geneva.
7 Napierville.
8 Yorkville.
9 Morris.
10 Keithsburg.
11 Oquawka.
12 Knoxville.
13 Toulon.
14 Hennepin.
15 Lacon.
16 Metamora.
17 Pontiac.
18 Carthage.
19 Macomb.
20 Lewiston.
21 Havana.
22 Lincoln.
23 Clinton.
24 Monticello.
25 Paxton.
26 Mt. Sterling.
27 Rushville.
28 Beardstown.
29 Petersburg.
30 Pittsfield.
31 Winchester.
32 Taylorsville.
33 Sullivan.
34 Tuscola.
35 Carrollton.
36 Hardin.
37 Jerseyville.
38 Hillsboro.
39 Windsor.
40 Greenup.
41 Marshall.
42 Edwardsville.
43 Greenville.
44 Vandalia.
45 Effingham.
46 Newton.
47 Robinson.
48 Carlyle.
49 Salem.
50 Lewisville.
51 Waterloo.
52 Nashville.
53 Mt. Vernon.
54 Fairfield.
55 Albion.
56 Mt. Carmel.
57 Chester.
58 Pinckneyville.
59 Benton.
60 McLeansboro.
61 Murphysboro.
62 Marion.
63 Harrisburg.
64 Jonesboro.
65 Vienna.
66 Golconda.
67 Elizabeth.
68 Thebes.
69 Caledonia.
70 Metropolis.

INDIANA.
1 Crown Point.
2 Valparaiso.
3 La Grange.
4 Angola.
5 Knox.
6 Plymouth.
7 Warsaw.
8 Albion.
9 Auburn.
10 Rensselaer.
11 Winamac.
12 Rochester.
13 Columbia.
14 Kent.
15 Monticello.
16 Wabash.
17 Huntington.
18 Bluffton.
19 Decatur.
20 Oxford.
21 Delphi.
22 Kokomo.
23 Marion.
24 Hartford.
25 Portland.
26 Williamsport.
27 Covington.
28 Frankfort.
29 Tipton.
30 Anderson.
31 Winchester.
32 Lebanon.
33 Newcastle.
34 Newport.
35 Danville.
36 Greenfield.
37 Centreville.
38 Rockville.
39 BowlingGreen
40 Spencer.
41 Martinsville.
42 Franklin.
43 Shelbyville.
44 Rushville.
45 Connorsville
46 Liberty.
47 Sullivan.
48 Bloomfield.
49 Bloomington.
50 Nashville.
51 Greensburg.
52 Brookville.
53 Washington.
54 Dover Hill.
55 Bedford.
56 Brownstown.
57 Vernon.
58 Versailles.
59 Princeton.
60 Petersburg.
61 Jasper.
62 Paoli.
63 Salem.
64 Lexington.
65 Vevay.
66 Rising Sun.
67 Mt. Vernon.
68 Booneville.
69 Rockport.
70 Rome.
71 Fredonia.
72 Corydon.
73 Charleston.

IOWA.
1 Estherville.
2 Forest City.
3 Mitchell.
4 Cresco.
5 Waterman.
6 Emmettsburg.
7 Upper Grove.
8 Mason City.
9 Charles City.
10 Bradford.
11 West Union.
12 El Kader.
13 Spencer.
14 Rolfe.
15 Dakota.
16 Ontario.
17 Hampton.
18 Clarksville.
19 Melbourne.
20 Ida.
21 Sac City.
22 Lake City.
23 Webster.
24 Eldora.
25 Grundy.
26 Delhi.
27 Carrollton.
28 Jefferson.
29 Boonesboro.
30 Nevada.
31 Toledo.
32 Vinton.
33 Marion.
34 Bellevue.
35 Magnolia.
36 Harlan.
37 Exira.
38 Panora.
39 Adel.
40 Newton.
41 Montezuma.
42 Marengo.
43 Tipton.
44 Lewis.
45 Fontanelle.
46 Winterset.
47 Indianola.
48 Pella.
49 Sigourney.
50 Red Oak.
51 Quincy.
52 Afton.
53 Osceola.
54 Albia.
55 Fairfield.
56 Wapello.
57 Sidney.
58 Clarinda.
59 Bedford.
60 Mt. Ayr.
61 Leon.
62 Corydon.
63 Centreville.
64 Bloomfield.
65 Keosauqua.

KANSAS.
1 Kirwin.
2 Gaylord.
3 Belleville.
4 Washington.
5 Marysville.
6 Seneca.
7 Hiawatha.
8 Troy.
9 Beloit.
10 Clay Centre.
11 Louisville.
12 Holton.
13 Manhattan.
14 Oskaloosa.
15 Hays.
16 Russell.
17 Salina.
18 Abilene.
19 Alma.
20 Burlingame.
21 Marion.
22 Cottonw'd F'ls
23 Mound City.
24 Eldorado.
25 Eureka.
26 Sumner.
27 Winfield.
28 Fredonia.
29 Girard.
30 Peru.
31 Independence.

KENTUCKY.
1 Burlington.
2 Independence.
3 Alexandria.
4 Warsaw.
5 Williamstown
6 Falmouth.
7 Washington.
8 Clarksburg.
9 Greenupsb'g.
10 Bedford.
11 Carrollton.
12 Owenton.
13 Cynthiana.
14 Carlisle.
15 Flemingsburg
16 Morehead.
17 Grayson.
18 Catlettsburg.
19 Taylorsville.
20 Lawrenceburg
21 Georgetown.
22 Winchester.
23 Mt. Sterling.
24 Owingsville.
25 West Liberty.
26 Louisa.
27 Hardinsburg.
28 Bardstown.
29 Springfield.
30 Harrodsburg.
31 Nicholasville.
32 Richmond.
33 Irvine.
34 Salyersville.
35 Paintville.
36 Morganfield.
37 Dixon.
38 Calhoun.
39 Hartford.
40 Litchfield.
41 Hodgenville.
42 Lebanon.
43 Stanford.
44 Lancaster.
45 McKee.
46 Booneville.
47 Jackson.
48 Prestonburg.
49 Piketon.
50 Blandville.
51 Smithland.
52 Marion.
53 Eddyville.
54 Princeton.
55 Madisonville.
56 Greenville.
57 Morgantown.
58 Brownsville.
59 Munf'dsville.
60 Greensburg.
61 Liberty.
62 Mt. Vernon.
63 London.
64 Manchester.
65 Hazard.
66 Whitesburg.
67 Clinton.
68 Mayfield.
69 Murray.
70 Cadiz.
71 Elkton.
72 Franklin.
73 Scottsville.
74 Tompkinsv'le.
75 Glasgow.
76 Burkesville.
77 Columbia.
78 Jamestown.
79 Albany.
80 Monticello.
81 Somerset.
82 Williamsburg.
83 Barboursville.
84 Mt. Pleasant.

LOUISIANA.
1 Belleview.
2 Homer.
3 Farmersville.
4 Bastrop.
5 Providence.
6 Richmond.
7 Winnsboro'.
8 Columbia.
9 Vernon.
10 Winfield.
11 Mansfield.
12 Grand Cane.
13 Many.
14 Harrisonburg.
15 St. Joseph.
16 Vidalia.
17 Alexandria.
18 Marksville.
19 Point Coupee.
20 W. Baton R'ge
21 Lake Charles.
22 Houma.
23 Covington.
24 Frankliton.
25 Greensburg.
26 Clinton.
27 St. Francisv'le.

MAINE.
1 Machias.
2 Princeton.
3 Mattawamk'g.
4 Steuben.
5 Enfield.
6 Dover.
7 Frankfort.
8 Lincolnville.
9 Bloomfield.
10 Newport.
11 Readfield.
12 Jay.
13 Bethel.
14 Paris.

MARYLAND.
1 Westminster.
2 Elkton.
3 Chestertown.
4 Rockville.
5 Pr. Fred'kton.
6 Port Tobacco.
7 Cambridge.
8 Leonardtown.

MASS.
1 Brockton.
2 Chicopee.
3 Milford.
4 Weymouth.
5 Attleboro.
6 Woburn.
7 Westfield.

MICHIGAN.
1 Houghton.
2 Crossville.
3 Alpena.
4 Sherman.
5 Midland.
6 Vassar.
7 Lexington.
8 St. Johns.
9 Corunna.
10 Allegan.
11 Hastings.
12 Charlotte.
13 Mason.
14 Howell.
15 Pawpaw.
16 Berrien.
17 Cassopolis.
18 Centreville.

MINNESOTA.
1 OttertailCity.
2 Long Prairie.
3 Little Falls.
4 Hanover.
5 Brunswick.
6 Fortuna.
7 Twin Lakes.
8 Alexandria.
9 Winthrop.
10 Sauk Centre.
11 Wacab.
12 Cambridge.
13 Chengwatana.
14 Harrison.
15 Forest City.
16 Monticello.
17 Orono.
18 Taylor's Falls.
19 Glencoe.
20 Chaska.
21 Beaver.
22 New Ulm.
23 Le Sueur.
24 Watonwan.
25 Madelia.
26 Wilton.
27 Mantorville.
28 Oriachtown.
29 Jackson.
30 Fairmount.
31 Albert Lea.
32 Caledonia.

MISSISSIPPI.
1 Hernando.
2 Austin.
3 Ripley.
4 Fulton.
5 Pontotoc.
6 Coffeeville.
7 Charleston.
8 Friar's Point.
9 Bolivar.
10 McNutt.
11 Carrollton.
12 Greensboro'.
13 Starkville.
14 Kosciusko.
15 Lexington.
16 Macon.
17 De Kalb.
18 Philadelphia.
19 Yazoo City.
20 Brandon.
21 Raleigh.
22 Paulding.
23 Quitman.
24 Winchester.
25 Ellisville.
26 Williamsburg
27 Gallatin.
28 Port Gibson.
29 Woodville.
30 Liberty.
31 Holmesville.
32 Augusta.
33 Leakesville.
34 Mississip'i Cy.

MISSOURI.
1 Rockport.
2 Maryville.
3 Grant City.
4 Albany.
5 Bethany.
6 Princeton.
7 Unionville.
8 Lancaster.
9 Memphis.
10 Waterloo.
11 Maysville.
12 Trenton.
13 Milan.
14 Kirksville.
15 Edina.
16 Monticello.
17 Kingston.
18 Linneus.
19 Bloomington.
20 Shelbyville.
21 Palmyra.
22 Carrollton.
23 Huntsville.
24 Paris.
25 New London.
26 BowlingGreen
27 Marshall.
28 Fayette.
29 Columbia.
30 Fulton.
31 Danville.
32 Warrenton.
33 Troy.
34 Harrisonville.
35 Warsaw.
36 Versailles.
37 Tuscumbia.
38 Union.
39 Hillsboro.
40 Osceola.
41 Hermitage.
42 Linn Creek.
43 Rolla.
44 Steelville.
45 Farmington.
46 St. Genevieve.
47 Perryville.
48 Lamar.
49 Stockton.
50 Bolivar.
51 Buffalo.
52 Waynesville.
53 Salem.
54 Lesterville.
55 Fredericktown.
56 Jackson.
57 Carthage.
58 Mt. Vernon.
59 Ozark.
60 Marshfield.
61 Hartville.
62 Houston.
63 Eminence.
64 Greenville.
65 Benton.
66 Pineville.
67 Cassville.
68 Galena.
69 Forsyth.
70 Vera Cruz.
71 Gainesville.
72 West Plains.
73 Thomasville.
74 Poplar Bluff.
75 Bloomfield.
76 Charleston.
77 New Madrid.

NEBRASKA.
1 St. James.
2 Ponca.
3 Antelope.
4 Pierce.
5 La Porte.
6 Clinton.
7 West Point.
8 Tekama.
9 Schuyler.
10 Savannah.
11 Osceola.
12 Ashland.
13 Bellevue.
14 Orrville.
15 York.
16 Henry.
17 Pleasant Hill.
18 Hebron.
19 Tecumseh.
20 Pawnee City.

NEVADA.
1 Stillwater.
2 La Plata.
3 Aurora.
4 Hiko.
5 Humboldt.
6 Belmont.

NEW HAMP.
1 Gorham.
2 Milford.
3 Conway.
4 Moultonboro'.
5 Orford.
6 Wilton.
7 Franklin.
8 Farmington.
9 Amherst.

NEW JERSEY.
1 Newton.
2 Somerville.
3 Flemington.
4 Freehold.
5 Mt. Holly.
6 Tom's River.
7 Woodbury
8 Belvidere.
9 Salem.

NEW YORK.
1 Malone.
2 Mayville.
3 Little Valley.
4 Warsaw.
5 Sandy Hill.
6 Monticello.
7 Goshen.
8 Clarkstown.
9 Tarrytown.

N. CAROLINA.
1 Jefferson.
2 Boone.
3 Wilkesboro.
4 Danbury.
5 Winston.
6 Wentworth.
7 Yanceyville.
8 Graham.
9 Hillsboro'.
10 Roxboro'.
11 Louisburg.
12 Warrenton.
13 Halifax.
14 Winton.
15 Gatesville.
16 Elizabeth City
17 Williamston.
18 Hertford.
19 Greenville.
20 Onslow.
21 Kenansville.
22 Clinton.
23 Smithville.
24 Whitesville.
25 Lumberton.
26 Rockingham.
27 Carthage.
28 Pittsboro'.
29 Ashboro'.
30 Troy.
31 Wadesboro
32 Monroe.
33 Lexington.
34 Concord.
35 Mocksville.
36 Taylorsville.
37 Newton.
38 Lincolnton.
39 Dallas.
40 Shelby.
41 Rutherfordton
42 Columbus.
43 Waynesville.
44 Franklin.
45 Murphy.
46 Weldon.

OHIO.
1 Bryan.
2 Wauseon.
3 Perrysburg.
4 Clinton.
5 Bellevue.
6 Elyria.
7 Painesville.
8 Chardon.
9 Jefferson.
10 Defiance.
11 Napoleon.
12 Paulding.
13 Kalida.
14 Up. Sandusky.
15 Bucyrus.
16 Ashland.
17 Medina.
18 Ravenna.
19 Warren.
20 Canfield.
21 Van Wert.
22 Celina.
23 Wapakoneta.
24 Kenton.
25 Marion.
26 Mt. Gilead.
27 Mt. Vernon.
28 Wooster.
29 Millersburg.
30 New Phila.
31 Carrollton.
32 New Lisbon.
33 Sidney.
34 Bellefontaine.
35 Greenville.
36 Troy.
37 Marysville.
38 Coshocton.
39 Cadiz.
40 Eaton.
41 London.
42 Lancaster.
43 Somerset.
44 Cambridge.
45 St. Clairsville.
46 Lebanon.
47 Wilmington.
48 Washington.
49 Logan.
50 McConnellsv
51 Sarahsville.
52 Woodsfield.
53 McArthur.
54 Athens.
55 Batavia.
56 Georgetown.
57 West Union.
58 Piketon.
59 Jackson.
60 Gallipolis.

OREGON.
1 Forest Grove.
2 East Portland.
3 Jacksonville.
4 Astoria.
5 Ranier.

PENNSYLV'A.
1 Mercer.
2 Butler.
3 Brookville.
4 Emporium.
5 Coudersport.
6 Sunbury.
7 Milford.
8 Doylestown.
9 Lebanon.
10 McConnellsb'g
11 Ebensburg.
12 Indiana.
13 Kittanning.
14 Somerset.
15 Uniontown.
16 Waynesburg.
17 Lewisburg.

S. CAROLINA.
1 Anderson.
2 Spartanburg.
3 Yorkville.
4 Laurensville.
5 Unionville.
6 Chesterville.
7 Lancaster.
8 Chesterfield.
9 Marion.
10 Camden.
11 Edgefield.
12 Orangeburg.
13 Kingstree.
14 Barnwell.
15 Walterboro'.

TENN. (p. 102).

TEXAS.
1 Bonham.
2 Paris.
3 Mt. Pleasant.
4 Henderson.
5 Jasper.
6 Fort Duncan.
7 San Patricio.
8 Corvallitas.
9 RioGrandeCy.

UTAH.
1 Willard City.
2 Promontory.
3 Harmony.
4 Uintah.
5 Lehi.

VERMONT.
1 Newport.
2 Irasburg.
3 Hyde Park.
4 Chelsea.
5 Fayetteville.
6 Shaftesbury.

VIRGINIA.
1 Woodstock.
2 Front Royal.
3 Warrenton.
4 Washington.
5 Luray.
6 Culpepper.
7 Madison.
8 Stanardville.
9 Spottsylvania.
10 Monterey.
11 Bowl'g Green.
12 Louisa.
13 Tappahan'ock
14 Hanover.
15 Heathsville.
16 Lancaster.
17 Saluda.
18 Smithfield.
19 Surry.
20 Prince George
21 Suffolk.
22 Jerusalem.
23 Sussex.
24 Goochland.
25 Powhatan.
26 Amelia.
27 Nottoway.
28 Lunenburg.
29 Boydton.
30 Charlotte.
31 Ap'mat'x C.H.
32 Lovingston.
33 Amherst.
34 Campbell.
35 Pittsylvania.
36 Martinsville.
37 Rocky Mount.
38 Fincastle.
39 Newcastle.
40 Christiansb'g.
41 Parisburg.
42 Patrick
43 Hillsville.
44 Bland.
45 Marion.
46 Buchanan.
47 Gladesville.
48 Estillville.

WASH. TER.
1 Montesano.
2 Oysterville.
3 Port Madison.
4 Cathlamet.
5 SnohomishCy.

W. VIRGINIA.
1 St. Mary's.
2 Clarksburg.
3 New Creek.
4 Bath.
5 Romney.
6 Grant.
7 Moorefield.
8 Franklin.
9 St. George.
10 Harrisville.
11 Grantsville.
12 Spencer.
13 Jackson.
14 Clay.
15 Hamlin.
16 Wayne.
17 Ballardsville.
18 Perrysville.

WISCONSIN.
1 Grantsburg.
2 St.Croix Falls.
3 Prescott.
4 Menomonee.
5 Neillsville.
6 Wausau.
7 Sturgeon Bay.
8 Pepin.
9 Fountain City.
10 Galesville.
11 Black R.Falls.
12 Grand Rapids.
13 Plover.
14 Mukwa.
15 Kewaunee.
16 Viroqua.
17 New Lisbon.
18 Ft. Atkinson.
19 Montello.
20 Wautoma.
21 Chilton.
22 Richland.
23 Baraboo.
24 Juneau.
25 West Bend.
26 Ozaukee.
27 Lancaster.
28 Jefferson.
29 Elkhorn.
30 Waupaca.
31 Mauston.

CHART SHOWING THE COMPARATIVE AREAS OF STATES, COUNTRIES, Etc.

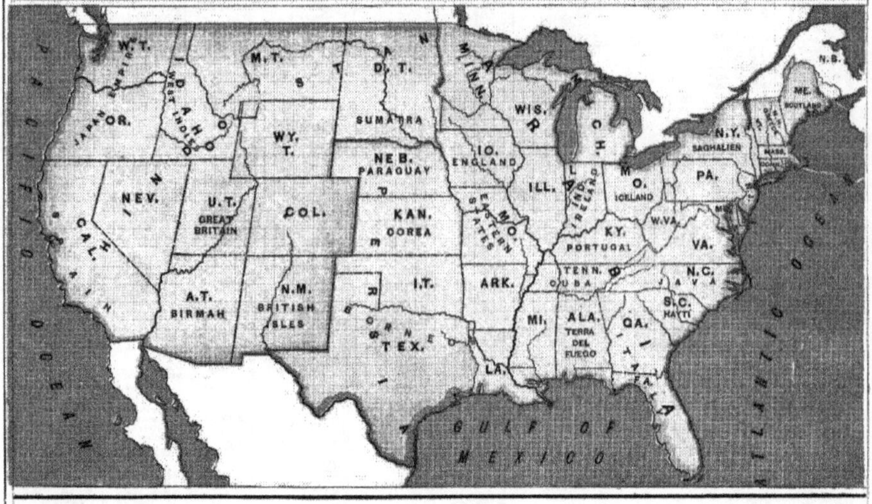

The Countries and Islands have areas equal, or nearly equal, to those of the States in which their names appear. For example, the area of Iowa is about the same as that of England.
The portion of the map which is colored yellow represents the area of Arabia; that colored red, Hindoostan; green, Persia.

What three countries in Asia taken together cover an area nearly equal to that of the United States, without Alaska?

How many States together comprise an area equal to that of Arabia? *Ans.* 31.

What States are included in this area? *Ans. All those between the Atlantic Ocean and the Mississippi River, besides the five States which are situated on the west side of that river.*

What State is in the north-eastern part of the region which is here compared with Arabia? In the south-western part? North-western? South-eastern part?

What rivers flow through that part of the United States?

What States and Territories together comprise an area equal to that of Hindoostan?

What mountains extend through that part of the United States?

What are included in the region whose area is equal to that of Persia?

Bound that part of the United States which has the same area as Arabia. Persia. Hindoostan.

Which is the largest State in the Union?

What country in Europe has about the same area as California? Kentucky? Maine? Iowa? Indiana?

What country is the same size as New Hampshire and Vermont combined? What country is but little larger than Georgia and Florida combined?

What country in South America is nearly as large as Nebraska?

What empire has an area nearly equal to that of Oregon and Washington combined?

What islands comprise an area equal to that of Idaho? Of New Mexico?

What island has an area nearly the same as that of the State of New York? Ohio? Alabama? North Carolina? Dakota? Tennessee? South Carolina? Utah? Texas?

What six States have the same area as Missouri?

What country in Asia contains the same number of square miles as Kansas?

What country in Asia is the same size as Arizona?

What State or what Territory has an area equal to that of

Great Britain?	Scotland?	Corea?	Saghalien?
Spain?	Ireland?	Paraguay?	The Eastern or New England States?
Portugal?	Iceland?	Terra del Fuego?	West Indies?
The British Isles?	Birmah?	Sumatra?	Cuba?
England?	Hayti?	Java?	

What part of the United States has the same area as Italy? Greece? Japan Empire? Borneo?

AREAS IN SQUARE MILES.

	Sq. miles.		Sq. miles.		Sq. miles.
Maine	35,000	Tennessee	45,600	Missouri	65,000
Scotland	31,324	Cuba	43,800	Eastern States	68,000
Vermont & N. Hamp.	19,000	N. Carolina	50,704	California	189,000
Greece & Ionian Is.	19,000	Java	51,000	Spain	195,000
Ohio	39,964	S. Carolina	34,000	Montana	143,000
Iceland	35,000	Liberia	26,000	Prussia	135,000
Indiana	33,809	Georgia & Florida	109,000	Hindoostan	1,500,000
Ireland	32,500	Italy	114,000	Arabia	1,000,000
				Persia	500,000
England	50,922	Texas	274,000	Hind., Ara. & Persia	3,000,000
Iowa	55,000	Borneo	284,000	Australia	3,000,000
Kentucky	37,680	New Mexico	121,000	United States, without Alaska	3,000,000
Portugal	36,000	British Isles	128,000		

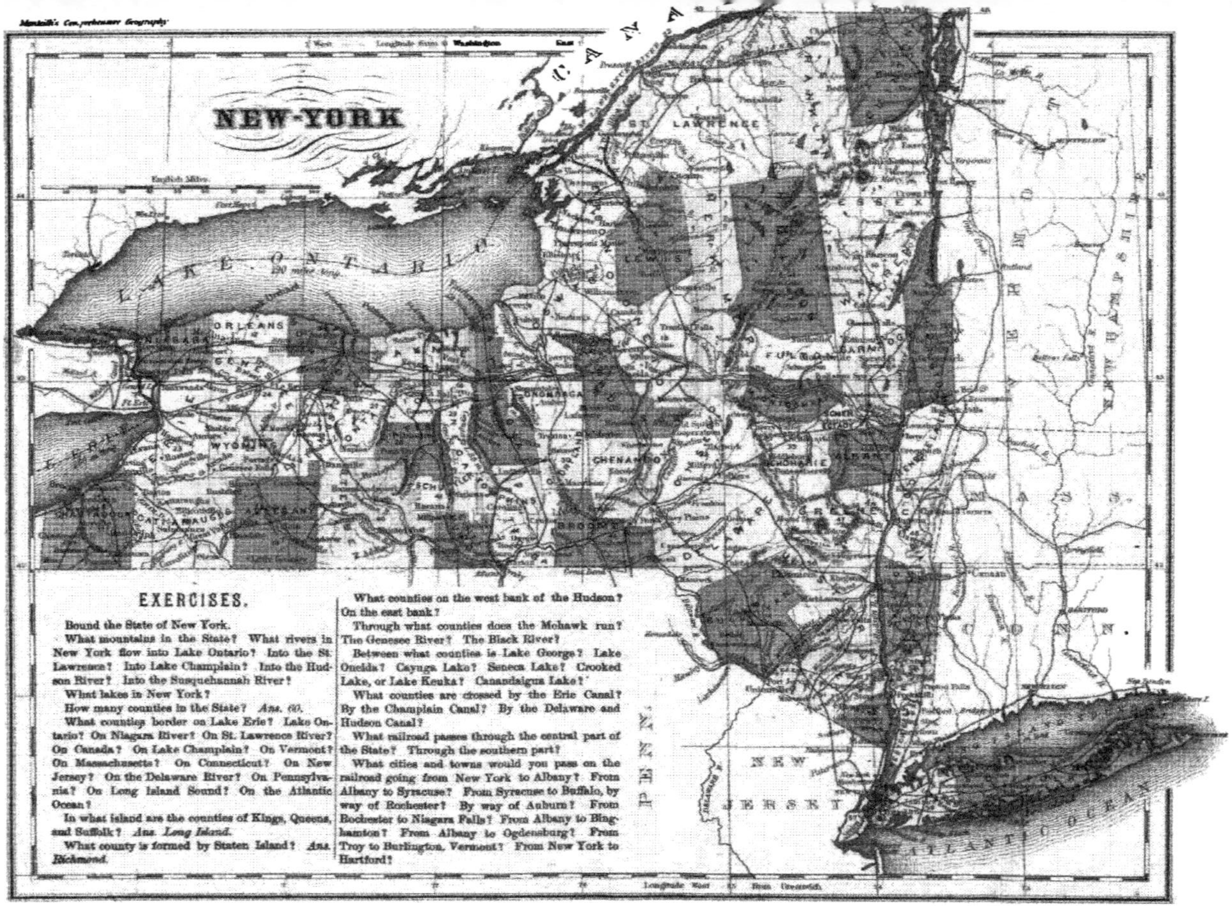

EXERCISES.

Bound the State of New York.

What mountains in the State? What rivers in New York flow into Lake Ontario? Into the St Lawrence? Into Lake Champlain? Into the Hudson River? Into the Susquehannah River?

What lakes in New York?

How many counties in the State? *Ans. 60.*

What counties border on Lake Erie? Lake Ontario? On Niagara River? On St. Lawrence River? On Canada? On Lake Champlain? On Vermont? On Massachusetts? On Connecticut? On New Jersey? On the Delaware River? On Pennsylvania? On Long Island Sound? On the Atlantic Ocean?

In what island are the counties of Kings, Queens, and Suffolk? *Ans. Long Island.*

What county is formed by Staten Island? *Ans. Richmond.*

What counties on the west bank of the Hudson? On the east bank?

Through what counties does the Mohawk run? The Genesee River? The Black River?

Between what counties is Lake George? Lake Oneida? Cayuga Lake? Seneca Lake? Crooked Lake, or Lake Keuka? Canandaigua Lake?

What counties are crossed by the Erie Canal? By the Champlain Canal? By the Delaware and Hudson Canal?

What railroad passes through the central part of the State? Through the southern part?

What cities and towns would you pass on the railroad going from New York to Albany? From Albany to Syracuse? From Syracuse to Buffalo, by way of Rochester? By way of Auburn? From Rochester to Niagara Falls? From Albany to Binghamton? From Albany to Ogdensburg? From Troy to Burlington, Vermont? From New York to Hartford?

Appendix to MONTEITH'S Geography.

Monteith's Globe.

The Earth's Surface represented on Segments from which every pupil can make a Globe.

DIRECTIONS.

Cut the segments apart and paste them on a ball three inches in diameter: the segments, while damp with the paste, can be easily adapted to the curve of the ball by slightly rubbing and stretching them.

If a ball cannot be obtained, paste the segments side by side on a slip of stiff paper, one quarter of an inch in width, placing the Equator over the middle of the slip. Unite the ends of the slip, and then paste the points of the segments on a small circular piece of paper at the Poles.

(Should the Map be spoiled in following out these directions to make a globe, a duplicate copy can be obtained of Messrs. A. S. BARNES & Co., New York, who will mail one to any part of the country on receipt of 50 cents.)

EXERCISES ON THE GLOBE.

Point to the Equator,— the North Pole, — the South Pole,—to the Meridians.

How many meridians are drawn on this and most other globes?

In what time does the Earth revolve on its axis?

How many degrees in every Circle?—In every Semi-circle?—In a quarter of a Circle?

How many degrees from the North to the South Pole? — From the Equator to either Pole?

How many degrees from one meridian to another on this globe?

Point to the Parallels of latitude, — to the Tropic of Cancer, — Tropic of Capricorn, — the Arctic Circle,—the Antarctic Circle.

What countries are in the Torrid Zone? — North Temperate Zone? — South Temperate Zone?

Place one end of your pencil on the globe, in the middle of the United States, so that it will point to the centre of the globe; that is your position when you stand up. Place a pencil in like manner in various points of the Earth's surface, and you will see the positions of the inhabitants when standing erect.

Place one pencil thus on Spain and another on New Zealand, each pencil representing a man; when they both look or point downward they look or point directly toward each other. The New Zealanders are the antipodes of the Spaniards.

Downward is always *toward*, and Upward *from* the Earth's centre.

What part of the Earth's surface is diametrically opposite you?

In what direction do all meridians run? Turn to the various maps in your Geography, and observe that while the meridians do not run parallel with the sides of the maps, their direction is north and south.

At what two points on the globe do all meridians end?

The teacher may here exercise the class in latitudes and longitudes.

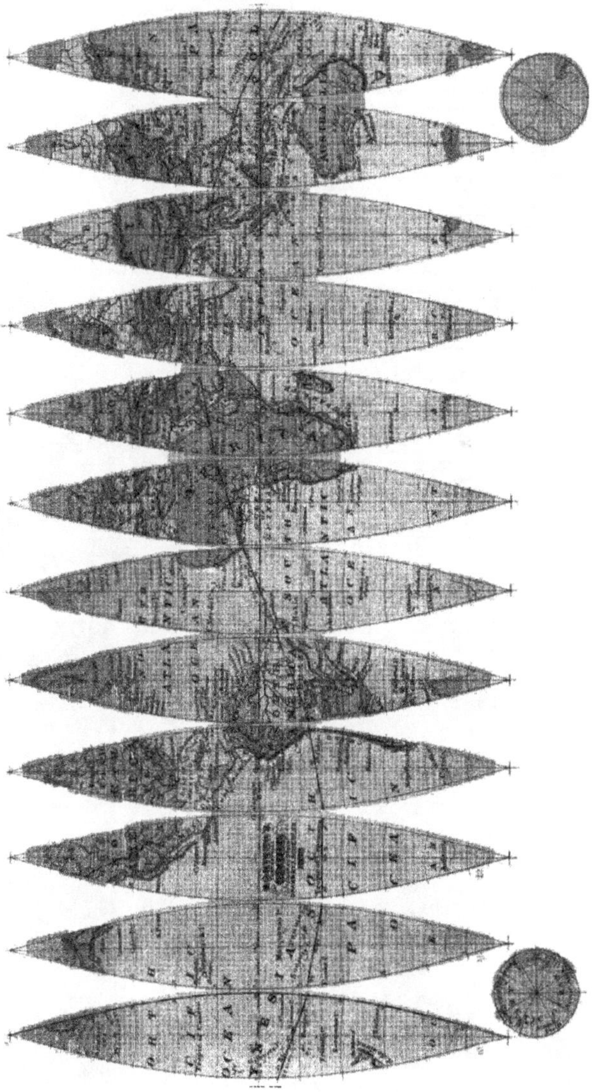

MONTEITH'S GLOBES.

Published by **E. Steiger**, ***New York;*** for sale by **A. S. Barnes & Co.**, ***New York*** and ***Chicago***, and by Booksellers and Stationers generally.

TERRESTRIAL GLOBES.

Beautifully printed in colors, the water blue, the ocean currents white, indicating the principal lines of Ocean Steam Communication, and Submarine Telegraph Cables.

(The prices within brackets [ ] denote the extra cost of packing which will be charged to the purchaser.)

V A. ***The twelve-inch Globe.*** Complete. On low bronzed frame, with horizon, meridian, hour-circle, and quadrant. 25.00 [2.50]

V B. ***The twelve-inch Globe.*** With full Meridian. On bronzed stand, with full meridian, and inclined axis. 18.00 [2.00]

V C. ***The twelve-inch Globe.*** Plain. On low bronzed stand, with inclined axis. 15.00 [2.00]

V G. ***The twelve-inch Globe.*** On bronzed hinged bracket. 10.00 [1.50]

V L. ***The twelve-inch Suspended Globe.*** With meridian. 10.00 [1.25]

VI A. ***The nine-inch Terrestrial Globe.*** Complete. On low iron frame, with horizon, meridian, hour-circle, and quadrant. 16.00 [1.80]

VI B. ***The nine-inch Terrestrial Globe.*** With full meridian. On low iron stand, with full meridian, and inclined axis. 12.00 [1.20]

VI C. ***The nine-inch Terrestrial Globe.*** Plain. On plain iron stand, with inclined axis. 9.00 [1.00]

VI G. ***The nine-inch Terrestrial Globe.*** On bronzed hinged bracket. 6.00 [0.60]

VI L. ***The nine-inch Suspended Globe.*** With meridian. 8.00 [0.60]

VII A. ***The six-inch Globe.*** Complete. On low iron frame, with horizon, meridian, and hour-circle. 10.00 [0.60]

VII B. ***The six-inch Globe.*** With full meridian. On low iron stand, with full meridian, and inclined axis. 5.00 [0.50]

VII C. ***The six-inch Globe.*** Plain. On low iron stand, with inclined axis. 4.00 [0.40]

VII D. ***The six-inch Globe.*** In Paper Box. (The Globe, when used, to be put on the top of the Box.) ***The Public School Globe.*** 3.00 [0.40]

VII G. ***The six-inch Globe.*** On bronzed hinged bracket. 4.00 [0.40]

VII L. ***The six-inch Suspended Globe.*** With meridian. 6.00 [0.40]

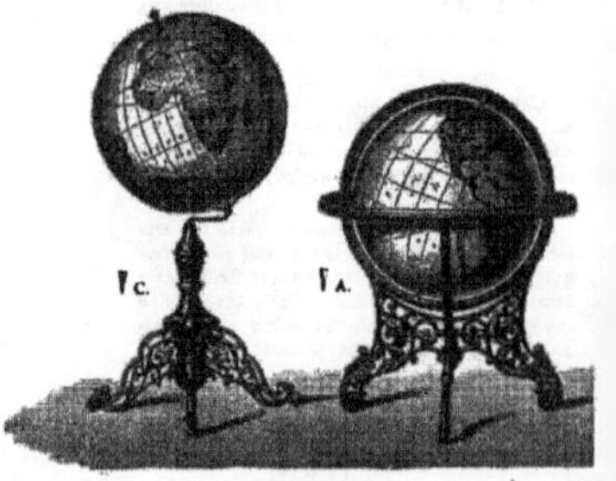

V C. V A.

Hemisphere

TERRESTRIAL GLOBES, of 6 Inches Diameter.

The two styles of Hemisphere Globe mentioned below, are a most important addition to cheap school-apparatus. In both, the arrangement at once shows the learner why the lines on a map must be curved; how impossible it is to depict perfectly any part of the Globe on a flat map, or to represent on such a map, in their correct form, and in complete unity, countries and seas in their natural proportions, position, distances, etc. For it is clear that a sphere, or part of a sphere, cannot be accurately represented upon a flat surface The juxtaposition of the Hemisphere Globe with the Planisphere Map, proves this to evidence, inasmuch as the comparison of the two shows very distinctly how distorted and disarranged all the parts of the Earth appear upon the Planisphere Map.

VII E. ***The Hand Hemisphere Globe.*** With hinge. 3.00 [0.40]

VII F. ***The Wall Hemisphere Globe.—Planisphere Maps*** and ***Hemisphere Globes*** combined. Mounted on pasteboard Patented October 21st, 1875.) 3.75 [0.40]

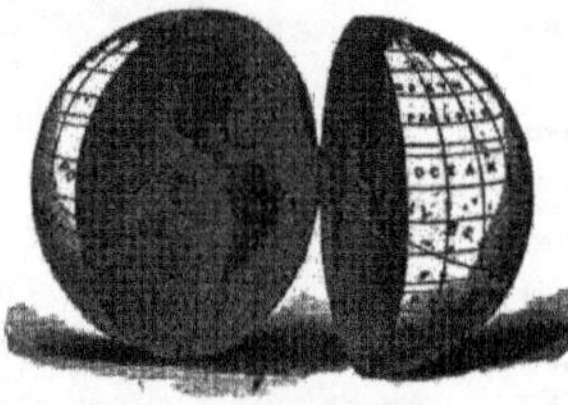

V G.

BRACKET GLOBES.

This is an entirely novel and most advantageous method of mounting the Globe.

For Common Schools, in which the teacher is not, as a rule, engaged in problems requiring the Globe to have stand, meridian, horizon, etc., this is the best kind.

The teacher needs an *inexpensive* Globe which can be placed beyond the reach of the scholars and the danger of accidental damage, can be readily taken down and handed 'round the class, and as quickly put back in its proper place. All these requirements will be found fully met in the Bracket Globe, of which *five* different sizes are offered.

SUSPENDED GLOBES.

This style will be found very serviceable wherever floor or table space cannot conveniently be spared for a Globe. The very low price at which the several sizes are offered, is another point in their favor.— Each Globe is provided with a cast-iron nickel-plated full meridian, 2 bird-cage pulleys, 1 pin and 6 yards of strong cord.

CELESTIAL GLOBES.—SLATED GLOBES.—TELLURIANS.

www.ingramcontent.com/pod-product-compliance
Lightning Source LLC
Chambersburg PA
CBHW021826190326
41518CB00007B/757
* 9 7 8 3 7 4 2 8 0 7 3 8 0 *